PRAISE FOR
THE SHAPE OF DATA

"The title says it all. Data is bound by many complex relationships not easily shown in our two-dimensional, spreadsheet-filled world. *The Shape of Data* walks you through this richer view and illustrates how to put it into practice."

—STEPHANIE THOMPSON, DATA SCIENTIST AND SPEAKER

"*The Shape of Data* is a novel perspective and phenomenal achievement in the application of geometry to the field of machine learning. It is expansive in scope and contains loads of concrete examples and coding tips for practical implementations, as well as extremely lucid, concise writing to unpack the concepts. Even as a more veteran data scientist who has been in the industry for years now, having read this book I've come away with a deeper connection to and new understanding of my field."

—KURT SCHUEPFER, PHD, MCDONALD'S CORPORATION

"*The Shape of Data* is a great source for the application of topology and geometry in data science. Topology and geometry advance the field of machine learning on unstructured data, and *The Shape of Data* does a great job introducing new readers to the subject."

—UCHENNA "IKE" CHUKWU, SENIOR QUANTUM DEVELOPER

THE SHAPE OF DATA

Geometry-Based Machine Learning and Data Analysis in R

by Colleen M. Farrelly
and Yaé Ulrich Gaba

no starch press®

San Francisco

THE SHAPE OF DATA. Copyright © 2023 by Colleen M. Farrelly and Yaé Ulrich Gaba.

All rights reserved. No part of this work may be reproduced or transmitted in any form or by any means, electronic or mechanical, including photocopying, recording, or by any information storage or retrieval system, without the prior written permission of the copyright owner and the publisher.

Printed in the United States of America

First printing

27 26 25 24 23   1 2 3 4 5

ISBN-13: 978-1-7185-0308-3 (print)
ISBN-13: 978-1-7185-0309-0 (ebook)

Publisher: William Pollock
Managing Editor: Jill Franklin
Production Manager: Sabrina Plomitallo-González
Production Editor: Sydney Cromwell
Developmental Editor: Alex Freed
Cover Illustrator: Gina Redman
Interior Design: Octopod Studios
Technical Reviewer: Franck Kalala Mutombo
Copyeditor: Kim Wimpsett
Compositor: Jeff Lytle, Happenstance Type-O-Rama
Proofreader: Scout Festa
Indexer: BIM Creatives, LLC

For information on distribution, bulk sales, corporate sales, or translations, please contact No Starch Press® directly at info@nostarch.com or:

No Starch Press, Inc.
245 8th Street, San Francisco, CA 94103
phone: 1.415.863.9900
www.nostarch.com

Library of Congress Cataloging-in-Publication Data

```
Names: Farrelly, Colleen, author. | Gaba, Yaé Ulrich, author.
Title: The shape of data : network science, geometry-based machine learning, and topological data
    analysis in R / by Colleen M. Farrelly and Yaé Ulrich Gaba.
Description: San Francisco, CA : No Starch Press, [2023] | Includes bibliographical references.
Identifiers: LCCN 2022059967 (print) | LCCN 2022059968 (ebook) | ISBN 9781718503083 (paperback) |
    ISBN 9781718503090 (ebook)
Subjects: LCSH: Geometric programming. | Topology. | Machine learning. | System analysis--Data
    processing. | R (Computer program language)
Classification: LCC T57.825 .F37 2023  (print) | LCC T57.825  (ebook) | DDC 006.3/1--dc23/
    eng/20230301
LC record available at https://lccn.loc.gov/2022059967
LC ebook record available at https://lccn.loc.gov/2022059968
```

No Starch Press and the No Starch Press logo are registered trademarks of No Starch Press, Inc. Other product and company names mentioned herein may be the trademarks of their respective owners. Rather than use a trademark symbol with every occurrence of a trademarked name, we are using the names only in an editorial fashion and to the benefit of the trademark owner, with no intention of infringement of the trademark.

The information in this book is distributed on an "As Is" basis, without warranty. While every precaution has been taken in the preparation of this work, neither the authors nor No Starch Press, Inc. shall have any liability to any person or entity with respect to any loss or damage caused or alleged to be caused directly or indirectly by the information contained in it.

To my grandmother Irene Borree, who enjoyed our discussions about new technologies into her late nineties.
—Colleen M. Farrelly

To God Almighty, Yeshua Hamashiach, the Key to all treasures of wisdom and knowledge and the Waymaker.
To my beloved wife, Owolabi.
Thank you for believing.
To my parents, Prudence and Gilberte, and my siblings, Olayèmi, Boladé, and Olabissi.
To Jeff Sanders, the man I would call my "academic father."
—Yaé Ulrich Gaba

About the Authors

Colleen M. Farrelly is a senior data scientist whose academic and industry research has focused on topological data analysis, quantum machine learning, geometry-based machine learning, network science, hierarchical modeling, and natural language processing. Since graduating from the University of Miami with an MS in biostatistics, Colleen has worked as a data scientist in a variety of industries, including healthcare, consumer packaged goods, biotech, nuclear engineering, marketing, and education. Colleen often speaks at tech conferences, including PyData, SAS Global, WiDS, Data Science Africa, and DataScience SALON. When not working, Colleen can be found writing haibun/haiga or swimming.

Yaé Ulrich Gaba completed his doctoral studies at the University of Cape Town (UCT, South Africa) with a specialization in topology and is currently a research associate at Quantum Leap Africa (QLA, Rwanda). His research interests are computational geometry, applied algebraic topology (topological data analysis), and geometric machine learning (graph and point-cloud representation learning). His current focus lies in geometric methods in data analysis, and his work seeks to develop effective and theoretically justified algorithms for data and shape analysis using geometric and topological ideas and methods.

About the Technical Reviewer

Franck Kalala Mutombo is a professor of mathematics at Lubumbashi University and the former academic director of AIMS Senegal. He previously worked in a research position at University of Strathclyde and at AIMS South Africa in a joint appointment with the University of Cape Town. He holds a PhD in mathematical sciences from the University of Strathclyde, Glasgow, Scotland. He is an expert in the study and analysis of complex network structure and applications. His most recent research considers the impact of network structure on long-range interactions applied to epidemics, diffusion, and object clustering. His other research interests include differential geometry of manifolds, finite element methods for partial differential equations, and data science.

BRIEF CONTENTS

Foreword xv

Acknowledgments xvii

Introduction xix

Chapter 1: The Geometric Structure of Data 1

Chapter 2: The Geometric Structure of Networks 23

Chapter 3: Network Analysis 55

Chapter 4: Network Filtration 75

Chapter 5: Geometry in Data Science 95

Chapter 6: Newer Applications of Geometry in Machine Learning 131

Chapter 7: Tools for Topological Data Analysis 155

Chapter 8: Homotopy Algorithms 167

Chapter 9: Final Project: Analyzing Text Data 179

Chapter 10: Multicore and Quantum Computing 193

References 207

Index 221

CONTENTS IN DETAIL

FOREWORD — xv

ACKNOWLEDGMENTS — xvii

INTRODUCTION — xix
Who Is This Book For? xx
About This Book xxi
Downloading and Installing R xxi
Installing R Packages xxii
Getting Help with R xxii
Support for Python Users xxiii
Summary xxiii

1
THE GEOMETRIC STRUCTURE OF DATA — 1
Machine Learning Categories 2
    Supervised Learning 2
    Unsupervised Learning 3
    Matching Algorithms and Other Machine Learning 4
Structured Data 4
    The Geometry of Dummy Variables 5
    The Geometry of Numerical Spreadsheets 8
    The Geometry of Supervised Learning 10
Unstructured Data 17
    Network Data 17
    Image Data 18
    Text Data 20
Summary 21

2
THE GEOMETRIC STRUCTURE OF NETWORKS — 23
Introduction to Network Science 24
The Basics of Network Theory 25
    Directed Networks 26
    Networks in R 26
    Paths and Distance in a Network 28
Network Centrality Metrics 30
    The Degree of a Vertex 30
    The Closeness of a Vertex 31
    The Betweenness of a Vertex 32
    Eigenvector Centrality 33
    PageRank Centrality 34
    Katz Centrality 35
    Hub and Authority 35

Measuring Centrality in an Example Social Network 36
Additional Quantities of a Network 42
    The Diversity of a Vertex 42
    Triadic Closure 43
    The Efficiency and Eccentricity of a Vertex 44
    Forman–Ricci Curvature 45
Global Network Metrics 47
    The Interconnectivity of a Network. 48
    Spreading Processes on a Network 49
    Spectral Measures of a Network 49
Network Models for Real-World Behavior 51
    Erdös–Renyi Graphs 51
    Scale-Free Graphs 52
    Watts–Strogatz Graphs 52
Summary 54

3
NETWORK ANALYSIS   55

Using Network Data for Supervised Learning 56
    Making Predictions with Social Media Network Metrics 56
    Predicting Network Links in Social Media 58
Using Network Data for Unsupervised Learning 59
    Applying Clustering to the Social Media Dataset. 59
    Community Mining in a Network 61
Comparing Networks 64
Analyzing Spread Through Networks 66
    Tracking Disease Spread Between Towns 67
    Tracking Disease Spread Between Windsurfers 69
    Disrupting Communication and Disease Spread 72
Summary 74

4
NETWORK FILTRATION   75

Graph Filtration 76
From Graphs to Simplicial Complexes 81
Introduction to Homology 85
    Examples of Betti Numbers 85
    The Euler Characteristic 87
    Persistent Homology 88
Comparison of Networks with Persistent Homology 90
Summary 94

5
GEOMETRY IN DATA SCIENCE   95

Introduction to Distance Metrics in Data 96
Common Distance Metrics 98
    Simulating a Small Dataset 98
    Using Norm-Based Distance Metrics 99
    Comparing Diagrams, Shapes, and Probability Distributions 105

K-Nearest Neighbors with Metric Geometry 116
Manifold Learning 119
        Using Multidimensional Scaling 120
        Extending Multidimensional Scaling with Isomap 121
        Capturing Local Properties with Locally Linear Embedding 122
        Visualizing with t-Distributed Stochastic Neighbor Embedding 124
Fractals 125
Summary 129

6
NEWER APPLICATIONS OF GEOMETRY IN MACHINE LEARNING   131

Working with Nonlinear Spaces 132
        Introducing dgLARS 133
        Predicting Depression with dgLARS 136
        Predicting Credit Default with dgLARS 138
Applying Discrete Exterior Derivatives 140
Nonlinear Algebra in Machine Learning Algorithms 146
Comparing Choice Rankings with HodgeRank 149
Summary 153

7
TOOLS FOR TOPOLOGICAL DATA ANALYSIS   155

Finding Distinctive Groups with Unique Behavior 156
Validating Measurement Tools 159
Using the Mapper Algorithm for Subgroup Mining 161
        Stepping Through the Mapper Algorithm 162
        Using TDAmapper to Find Cluster Structures in Data 163
Summary 166

8
HOMOTOPY ALGORITHMS   167

Introducing Homotopy 167
Introducing Homotopy-Based Regression 169
Comparing Results on a Sample Dataset 174
Summary 177

9
FINAL PROJECT: ANALYZING TEXT DATA   179

Building a Natural Language Processing Pipeline 180
The Project: Analyzing Language in Poetry 181
        Tokenizing Text Data 183
        Tagging Parts of Speech 183
        Normalizing Vectors 184
Analyzing the Poem Dataset in R 184
Using Topology-Based NLP Tools 188
Summary 191

10
MULTICORE AND QUANTUM COMPUTING   193

Multicore Approaches to Topological Data Analysis 194
Quantum Computing Approaches 195
        Using the Qubit-Based Model 196
        Using the Qumodes-Based Model 197
        Using Quantum Network Algorithms 197
        Speeding Up Algorithms with Quantum Computing 199
        Using Image Classifiers on Quantum Computers 200
Summary 205

REFERENCES   207

INDEX   221

FOREWORD

The title of Colleen M. Farrelly and Yaé Ulrich Gaba's book, *The Shape of Data*, is as fitting and beautiful as the journey that the authors invite us to experience, as we discover the geometric shapes that paint the deeper meaning of our analytical data insights.

Enabling and combining common machine learning, data science, and statistical solutions, including the combinations of supervised/unsupervised or deep learning methods, by leveraging topological and geometric data analysis provides new insights into the underlying data problem. It reminds us of our responsibilities as data scientists, that with any algorithmic approach a certain data bias can greatly skew our expected results. As an example, the data scientist needs to understand the underlying data context well to avoid performing a two-dimensional Euclidean-based distance analysis when the underlying data needs to account for three-dimensional nuances, such as what a routing analysis would require when traveling the globe.

Throughout the book's mathematical data analytics tour, we encounter the origin of data analysis on structured data and the many seemingly unstructured data scenarios that can be turned into structured data, which enables standard machine learning algorithms to perform predictive and prescriptive analytical insights. As we ride through the valleys and peaks of our data, we learn to collect features along the way that become key inputs into other data layers, forming geometrical interpretations of varying unstructured data sources including network data, images, and text-based data. In addition, Farrelly and Gaba are masterful in detailing the foundational and advanced concepts supported by the well-defined examples in both R and Python, available for download from their book's web page.

Throughout my opportunities to collaborate with Farrelly and Gaba on several exciting projects over the past years, I always hoped for a book to emerge that would explain as clearly and eloquently as *The Shape of Data* does the evolution of the topological data analysis space all the way to leveraging distributed and quantum computing solutions.

During my days as a CTO at Cypher Genomics, Farrelly was leading our initiatives in genomic data analytics. She immediately inspired me with her keen understanding of how best to establish correlations between disease ontologies versus symptom ontologies, while also using simulations to understand the implications of missing links in the map. Farrelly's pragmatic approach helped us successfully resolve critical issues by creating an algorithm that mapped across gene, symptom, and disease ontologies in order to predict disease from gene or symptom data. Her focus on topology-based network mining for diagnostics helped us define the underlying data network shape, properties, and link distributions using graph summaries and statistical testing. Our combined efforts around ontology mapping, graph-based prediction, and network mining and decomposition resulted in critical data network discoveries related to metabolomics, proteomics, gene regulatory networks, patient similarity networks, and variable correlation networks.

From our joint genomics and related life sciences analytics days to our most recent quantum computing initiatives, Farrelly and Gaba have consistently demonstrated a strong passion and unique understanding of all the related complexities and how to apply their insights to several everyday problems. Joining them on their shape of data journey will be valuable time spent as you embark on a well-scripted adventure of R and Python algorithms that solve general or niche problems in machine learning and data analysis using geometric patterns to help shape the desired results.

This book will be relevant and captivating to beginners and devoted experts alike. First-time travelers will find it easy to dive into algorithm examples designed for analyzing network data, including social and geographic networks, as well as local and global metrics, to understand network structure and the role of individuals in the network. The discussion covers clustering methods developed for use on network data, link prediction algorithms to suggest new edges in a network, and tools for understanding how, for example, processes or epidemics spread through networks.

Advanced readers will find it intriguing to dive into recently developing topics such as replacing linear algebra with nonlinear algebra in machine learning algorithms and exterior calculus to quantity needs in disaster planning. *The Shape of Data* has made me want to roll up my sleeves and dive into many new challenges, because I feel as well equipped as Lara Croft in *Tomb Raider* thanks to Farrelly's tremendous treasure map and deeply insightful exploration work. Could there be a hidden bond or "hidden layer" between them?

Michael Giske
Technology executive, global CIO of B-ON,
and chairman of Inomo Technologies

ACKNOWLEDGMENTS

I, Colleen, would like to thank my parents, John and Nancy, and my grandmother Irene for their support while I was writing this book and for encouraging me to play with mathematics when I was young.

I would also like to thank Justin Moeller for the sports and art conversations that led me into data science as a career, as well as his and Christy Moeller's support over the long course of writing this book, and Ross Eggebeen, Mark Mayor, Matt Mayor, and Malori Mayor for their ongoing support with this project and other writing endeavors over the years.

My career in this field and this book would not have been possible without the support of Cynthia DeJong, John Pustejovsky, Kathleen Karrer, Dan Feaster, Willie Prado, Richard Schoen, and Ken Baker during my educational years, particularly my transition from the medical/social sciences to mathematics during medical school. I'm grateful for the support of Jay Wigdale and Michael Giske over the course of my career, as well as the support from many friends and colleagues, including Peter Wittek, Diana Kachan, Recinda Sherman, Natashia Lewis, Louis Fendji, Luke Robinson, Joseph Fustero, Uchenna Chukwu, Jay and Jenny Rooney, and Christine and Junwen Lin.

This book would not have been possible without our editor, Alex Freed; our managing editor, Jill Franklin; and our technical reviewer, Franck Kalala Mutombo. We both would also like to acknowledge the contributions of

Bastian Rieck and Noah Giansiracusa. We are grateful for the support of No Starch Press's marketing team, particularly Briana Blackwell in publicizing our speaking engagements.

We are also grateful to R, which provided open source packages and graphics generated with code, as well as Microsoft PowerPoint, which was used with permission to generate the additional images in this book. We would also like to thank NightCafe for providing a platform to generate images and granting full rights to creators.

No achievement in life is without the help of many known and unknown individuals. I, Yaé, would like to thank just a few who made this work possible.

To my wife, Owolabi, for your unwavering support. To Colleen Farrelly, for initiating this venture and taking me along. To Franck Kalala, my senior colleague and friend, for his excellent reviewer skills.

To my friends and colleagues: Collins A. Agyingi, David S. Attipoe, Rock S. Koffi, Evans D. Ocansey, Michael Kateregga, Mamana Mbiyavanga, Jordan F. Masakuna and Gershom Buri for the care. To Jan Groenewald and the entire AIMS-NEI family.

To my spiritual fathers and mentors, Pst Dieudonné Kantu and the entire SONRISE family, Pst Daniel Mukanya, and Pst Magloire N. Kunantu, whose leadership inpired my own.

INTRODUCTION

The first time I, Colleen, confronted my own hesitancy with math was when geometry provided a solution to an art class problem I faced: translating a flat painting onto a curved vase. Straight lines from my friend's canvas didn't behave the same way on the curved vase. Distances between points on the painting grew or shrank with the curvature. We'd stumbled upon the differences between the geometry we'd learned in class (where geometry behaved like the canvas painting) and the geometry of real-world objects like the vase. Real-world data often behaves more like the vase than the canvas painting.

As an industry data scientist, I've worked with many non-data-science professionals who want to learn new data science methods but either haven't encountered a lot of math or coding in their career path or have a lingering fear of math from prior educational experiences. Math-heavy papers without coding examples often limit the toolsets other professionals can use to solve important problems in their own fields.

Math is simply another language with which to understand the world around us; like any language, it's possible to learn. This book is focused on geometry, but it is not a math textbook. We avoid proofs, rarely use equations, and try to simplify the math behind the algorithms as much as possible to make these tools accessible to a wider audience. If you are more mathematically advanced and want the full mathematical theory, we provide references at the end of the book.

Geometry underlies every single machine learning algorithm and problem setup, and thousands of geometry-based algorithms exist today. This book focuses on a few dozen algorithms in use now, with preference given to those with packages to implement them in R. If you want to understand how geometry relates to algorithms, how to implement geometry-based algorithms with code, or how to think about problems you encounter through the lens of geometry, keep reading.

Who Is This Book For?

Though this book is for anyone anywhere who wants a hands-on guide to network science, geometry-based aspects of machine learning, and topology-based algorithms, some background in statistics, machine learning, and a programming language (R or Python, ideally) will be helpful. This book was designed for the following:

- Healthcare professionals working with small sets of patient data
- Math students looking for an applied side of what they're learning
- Small-business owners who want to use their data to drive sales
- Physicists or chemists interested in using topological data analysis for a research project
- Curious sociologists who are wary of proof-based texts
- Statisticians or data scientists looking to beef up their toolsets
- Educators looking for practical examples to show their students
- Engineers branching out into machine learning

We'll be surveying many areas of science and business in our examples and will cover dozens of algorithms shaping data science today. Each chapter will focus on the intuition behind the algorithms discussed and will provide examples of how to use those algorithms to solve a problem using the R programming language. While the book is written with examples presented in R, our downloadable repository (*https://nostarch.com/download/ShapeofData_PythonCode.zip*) includes R and Python code for examples where Python has an analogous function to support users of both languages. Feel free to skip around to sections most relevant to your interests.

About This Book

This book starts with an introduction to geometry in machine learning. Topics relevant to geometry-based algorithms are built through a series of network science chapters that transition into metric geometry, geometry- and topology-based algorithms, and some newer implementations of these algorithms in natural language processing, distributed computing, and quantum computing. Here's a quick overview of the chapters in this book:

Chapter 1: The Geometric Structure of Data   Details how machine learning algorithms can be examined from a geometric perspective with examples from medical and image data

Chapter 2: The Geometric Structure of Networks   Introduces network data metrics, structure, and types through examples of social networks

Chapter 3: Network Analysis   Introduces supervised and unsupervised learning on network data, network-based clustering algorithms, comparisons of different networks, and disease spread across networks

Chapter 4: Network Filtration   Moves from network data to simplicial complex data, extends network metrics to higher-dimensional interactions, and introduces hole-counting in objects like networks

Chapter 5: Geometry in Data Science   Provides an overview on the curse of dimensionality, the role of distance metrics in machine learning, dimensionality reduction and data visualization, and applications to time series and probability distributions

Chapter 6: Newer Applications of Geometry in Machine Learning   Details several geometry-based algorithms, including supervised learning in educational data, geometry-based disaster planning, and activity preference ranking

Chapter 7: Tools for Topological Data Analysis   Focuses on topology-based unsupervised learning algorithms and their application to student data

Chapter 8: Homotopy Algorithms   Introduces an algorithm related to path planning and small data analysis

Chapter 9: Final Project: Analyzing Text Data   Focuses on a text dataset, a deep learning algorithm used in text embedding, and analytics of processed text data through algorithms from previous chapters

Chapter 10: Multicore and Quantum Computing   Dives into distributed computing solutions and quantum algorithms, including a quantum network science example and a quantum image analytics algorithm

Downloading and Installing R

We'll be using the R programming language in this book. R is easy to install and compatible with macOS, Linux, and Windows operating systems. You can choose the download for your system at *https://cloud.r-project.org*. You might be prompted to click a link for your geographic location (or a general cloud connection option). If you haven't installed R before, you can choose the first-time installation of the base, which is the first download option on the R for Windows page.

Once you click the first-time option, you should see a screen that will give you an option to download R for Windows.

After R downloads, you'll follow the installation instructions that your system provides as a prompt. This will vary slightly depending on the operating system. However, the installation guide will take you through the steps needed to set up R.

You may want to publish your projects or connect R with other open source projects, such as Python. RStudio provides a comfortable interface with options to connect R more easily with other platforms. You can find RStudio's download at *https://www.rstudio.com*. Once you download RStudio, simply follow your operating system's command prompts to install with the configurations that work best for your use case.

Installing R Packages

R has several options for installing new packages on your system. The command line option is probably the easiest. You'll use the install.packages ("*package_name*") option, where *package_name* is the name of the package you want to install, such as install.packages("mboost") to install the mboost package. From there, you may be asked to choose your geographic location for the download. The package will then download (and download any package dependencies that are not already on your machine).

You can also use your graphical user interface (GUI) to install a package. This might be preferable if you want to browse available packages rather than install just one specific package to meet your needs. You can select **Install package(s)** from the Packages menu option after you launch R on your machine.

You'll be prompted to select your location, and the installation will happen as it would with the command line option for package installation.

Getting Help with R

R has many useful features if you need help with a function or a package in your code. The help() function allows you to get information about a function or package that you have installed in R. Adding the package name after the function (such as help(glmboost, "mboost") for help with the generalized linear modeling boosted regression function through the mboost package) will pull up information about a package not yet installed in your machine so that you can understand what the function does before deciding to install the new package. This is helpful if you're looking for something specific but not sure that what you're finding online is exactly what you need. In lieu of using the help() function, you can add a question mark before the function name (such as ?glmboost).

You can also browse for vignettes demonstrating how to use functions in a package using the command browseVignettes(), which will pull up vignettes for each package you have installed in R. If you want a vignette for a specific package, you can name that package like so: browseVignettes (package="mboost"). Many packages come with a good overview of how to apply the package's functions to an example dataset.

R has a broad user base, and internet searches or coding forums can provide additional resources for specific issues related to a package. There are also many good tutorials that overview the basic programming concepts and common functions in R. If you are less familiar with programming, you may want to go through a free tutorial on R programming or work with data in R before attempting the code in this book.

Because R is an evolving language with new packages added and removed regularly, we encourage you to keep up with developments via package websites and web searches. Packages that are discontinued can still be installed and used as legacy packages but require some caution, as they aren't updated by the package author. We'll see one of these in this book with an example of how to install a legacy package. Similarly, new packages are developed regularly, and you should find and use new packages in the field of geometry as they become available.

Support for Python Users

While this book presents examples in R code, our downloadable repository (*https://nostarch.com/download/ShapeofData_PythonCode.zip*) includes translations to Python packages and functions where possible. Most examples have a Python translation for Python users. However, some translations do not exist or include only the packages that do not install correctly on some operating systems. We encourage you to develop Python applications if they do not currently exist, and it is likely that more support in Python will become available for methods in this book (and methods developed in the future).

Summary

Mathematics can be intimidating for some people, but it forms the foundation of a lot of hot topics in machine learning and technology these days. Understanding the geometry behind the buzzwords will give you a deeper understanding of how the algorithms function and how they can be used to solve problems. You might even have some fun along the way.

We love helping people learn about geometry and machine learning. Feel free to connect with us or contact us on LinkedIn (*https://www.linkedin.com/in/colleenmfarrelly*).

This book will introduce you to geometry one step at a time. You'll probably have questions, wrestle with concepts, or try an example of your own based on what you read. Data science is a process. Getting help when you are stuck is a natural part of learning data science. Eventually, you'll find your own preferred methods of working through a problem you encounter.

Let's get started!

1

THE GEOMETRIC STRUCTURE OF DATA

You might wonder why you need data science methods rooted in topology and geometry when traditional data science methods are already so popular and powerful. The answer to this has two parts. First, data today comes in a variety of formats far more exotic than the usual spreadsheet, such as a social network or a text document. While such exotic data used to be referred to as *unstructured*, we now recognize that it often is structured, but with a more sophisticated geometric structure than a series of spreadsheets in a relational database. Topological and geometric data science allow us to work directly in these exotic realms and translate them into the more familiar realm of spreadsheets. Second, a relatively recent discovery suggests that geometry even lurks behind spreadsheet-structured data. With topological and geometric data science, we can harness the power of this hidden geometry.

We'll start this chapter with a quick refresher of the main concepts in traditional machine learning, discussing what it means for data to be structured and how this is typically used by machine learning algorithms. We'll then review supervised learning, overfitting, and the curse of dimensionality from a geometric perspective. Next, we'll preview a few other common types of data—network data, image data, and text data—and hint at how we can use their geometry in machine learning. If you're already familiar with traditional machine learning and the challenges of applying it to modern forms of data, you're welcome to skip to Chapter 2, where the technical content officially begins, although you may find the geometric perspectives of traditional machine learning topics offered in this chapter interesting regardless.

Machine Learning Categories

Many types of machine learning algorithms exist, and more are invented every day. It can be hard to keep track of all the latest developments, but it helps to think of machine learning algorithms as falling into a few basic categories.

Supervised Learning

Supervised learning algorithms generally aim to predict something, perhaps a treatment outcome under a new hospital protocol or the probability of a client leaving in the next six months. The variable we predict is called the *dependent variable* or *target*, and the variables used to predict it are called *independent variables* or *predictors*. When we're predicting a numerical variable (such as a symptom severity scale), this is called *regression*; when we're predicting a categorical variable (such as survived or died classes), this is called *classification*.

Some of the most popular supervised learning algorithms are *k*-nearest neighbors (*k*-NN), naive Bayes classifiers, support vector machines, random forests, gradient boosting, and neural networks. You don't need to know any of these topics to read this book, but it will help to be familiar with at least one regression method and one classification method, such as linear regression and logistic regression. That said, don't fret if you're not sure about them—this chapter will cover the concepts you need.

Each supervised learning algorithm is a specific type of function that has as many input variables as there are independent variables in the data. We think of this function as predicting the value of the dependent variable for any choice of values of the independent variables (see Figure 1-1). For linear regression with independent variables $x_1, x_2, \ldots, x_n$, this is a linear function: $f(x_1, x_2, \ldots, x_n) = a_1 x_1 + a_2 x_2 + \ldots +$. For other methods, it is a complicated nonlinear function. For many methods, this function is nonparametric, meaning we can compute it algorithmically but can't write it down with a formula.

Figure 1-1: The flow of independent variables, supervised machine learning algorithm (viewed as a function), and prediction of a dependent variable

Using a supervised learning algorithm usually involves splitting your data into two sets. There's *training* data, where the algorithm tries to adjust the parameters in the function so that the predicted values are as close as possible to the actual values. In the previous linear regression example, the parameters are the coefficients a_i. There's also *testing* data, where we measure how good a job the algorithm is doing. A *hyperparameter* is any parameter that the user must specify (as opposed to the parameters that are learned directly from the data during the training process). The k specifying number of nearest neighbors in k-NN is an example of a hyperparameter.

After a supervised learning algorithm has been trained, we can use it to make new predictions and to estimate the impact of each independent variable on the dependent variable (called *feature importance*). Feature importance is helpful for making intervention decisions. For instance, knowing which factors most influence patient death from an infectious disease can inform vaccination strategies when vaccine supplies are limited.

Unsupervised Learning

Unsupervised learning algorithms tend to focus on data exploration—for example, reducing the dimensionality of a dataset to better visualize it, finding how the data points are related to each other, or detecting anomalous data points. In unsupervised learning, there is no dependent variable—just independent variables. Accordingly, we don't split the data into training and testing sets; we just apply unsupervised methods to all the data. Applications of unsupervised learning include market segmentation and ancestry group visualization based on millions of genetic markers. Examples of unsupervised learning algorithms include k-means clustering, hierarchical clustering, and principal component analysis (PCA)—but again, you don't need to know any of these topics to read this book.

Unsupervised and supervised learning can be productively combined. For example, you might use unsupervised learning to find new questions about the data and use supervised learning to answer them. You might use unsupervised learning for dimension reduction as a preprocessing step to improve the performance of a supervised learning algorithm.

Matching Algorithms and Other Machine Learning

Another common application of machine learning involves *matching algorithms*, which compute the distance between points to find similar individuals within a dataset. These algorithms are commonly used to recommend a product to a user; they're also used in data integrity checks to make sure that sufficient data exists on certain populations before allowing a machine learning algorithm to create a model.

Data integrity is increasingly important with the rise of machine learning and artificial intelligence. If important subgroups within a population aren't captured well in the data, the algorithm will bias itself toward the majority groups. For example, if language recognition systems don't have sufficient data from African or Asian language groups, it's difficult for the system to learn human speech sounds unique to those language groups, such as Khoisan clicks or Mandarin tones. Matching algorithms are also used to try to tease out cause and effect from empirical evidence when a randomized controlled trial is not possible because they allow us to pair up similar participants as though they had been assigned to a treatment group and a placebo group.

We could mention many other types of algorithms, and, in practice, there tends to be overlap between algorithm categories. For instance, YouTube's recommendation algorithm uses *deep learning* (which involves machine learning based on neural networks with multiple "hidden layers") in both a supervised and an unsupervised way, as well as matching algorithms and another pillar of machine learning called *reinforcement learning* (where algorithms develop strategies on their own by exploring a real or simulated environment—beyond the reach of this book). However, the basic road map to machine learning provided earlier will guide us throughout this book.

Next, let's take a closer look at the format of the data these algorithms are expecting.

Structured Data

Machine learning algorithms, and data science and statistical methods more generally, typically operate on *structured data* (also called *tabular data*), which means a spreadsheet-type object (a data frame or matrix) in which the columns are the variables and the rows are the *data points* (also called *instances* or *observations*). These are usually stored in a relational database along with other structured data. The tabular structure is what allows us to talk about independent variables, dependent variables, and data points. A big focus of this book is how to deal with data that doesn't come presented in this nice format. But even with tabular data, a geometric perspective can be quite useful.

To start, let's dive into an example that will show how geometry can help us better understand and work with categorical variables in structured data.

The Geometry of Dummy Variables

Figure 1-2 shows part of a spreadsheet stored as a Microsoft Excel workbook. The last column here is Outcome, so we view that as the dependent variable; the other three columns are the independent variables. If we used this data for a supervised learning algorithm, it would be a regression task (since the dependent variable is numerical). The first independent variable is binary numerical (taking values 0 and 1), the second independent variable is discrete numerical (taking whole-number values), and the third independent variable is categorical (with three categories of gender). Some algorithms accept categorical variables, whereas others require all variables to be numerical. Unless the categories are ordered (such as survey data with values such as "very dissatisfied," "dissatisfied," "satisfied," and "very satisfied"), the way to convert a categorical variable to numerical is to replace it with a collection of binary *dummy variables*, which encode each category in a yes/no format represented by the values 1 and 0. In Figure 1-3, we've replaced the gender variable column with two dummy variable columns.

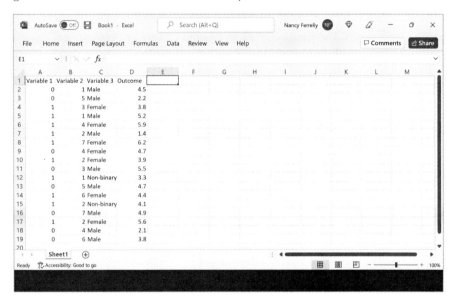

Figure 1-2: An example of structured data in a Microsoft Excel workbook

Even for such a common and simple process as this, geometric considerations help illuminate what is going on. If the values of a categorical variable are ordered, then we can convert them to a single numerical variable by placing the values along a one-dimensional axis in a way that reflects the order of these values. For example, the survey values "satisfied," "very satisfied," and "extremely satisfied" could be coded as 1, 2, and 3, or if you wanted the difference between "satisfied" and "very satisfied" to be smaller than the difference between "very satisfied" and "extremely satisfied," then you could code these as, say, 1, 2, and 4.

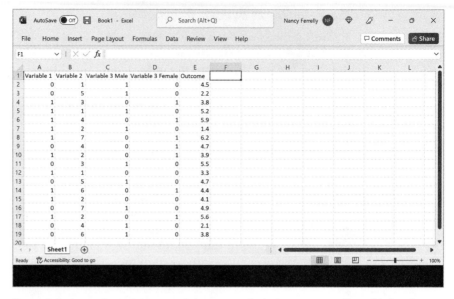

Figure 1-3: A transformed, structured dataset in which the categorical variable has been replaced with two binary dummy variables

If the categories are not ordered—such as Male, Female, and Non-Binary—we wouldn't want to force them all into one dimension because that would artificially impose an order on them and would make some of them closer than others (see Figure 1-4).

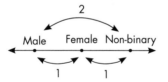

Figure 1-4: Placing the values of a categorical variable in a single dimension makes some of them closer than others.

Geometrically, we are creating new axes for the different categories when we create dummy variables. There are two ways of doing this. Sometimes, you'll see people use one dummy variable for each value of the categorical variable, whereas at other times you'll see people use dummy variables for all but one of the values (as we did in Figure 1-3). To understand the difference, let's take a look at our three-category gender variable.

Using three dummy variables places the categories as the vertices of an equilateral triangle: Male has coordinates (1,0,0), Female has coordinates (0,1,0), and Non-Binary has coordinates (0,0,1). This ensures the categories are all at the same distance from each other. Using only two dummy variables means Male has coordinates (1,0), Female has coordinates (0,1), and

Non-Binary has coordinates (0,0). This projects our equilateral triangle in three-dimensional space down to a right triangle in the two-dimensional plane, and in doing so it distorts the distances. Male and Female are now closer to Non-Binary than they are to each other, because they are separated by the length $\sqrt{2} \approx 1.4$ hypotenuse in this isosceles right triangle with side lengths 1 (see Figure 1-5). Consequently, some machine learning algorithms would mistakenly believe the categories Male and Female are more similar to the category Non-Binary than they are to each other.

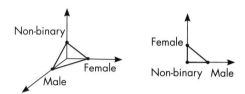

Figure 1-5: Two approaches to creating gender dummy variables. On the left, we have one axis for each category, which ensures the categories are equidistant. On the right, we use only two axes, which causes some categories to be closer than others.

So why are both dummy variable methods used? Using n dummy variables for an n-value categorical variable rather than $n-1$ leads to *multicollinearity*, which in statistical language is a correlation among the independent variables. The correlation here is that each of the dummy variables is completely and linearly determined by the others. Algebraically, this is a *linear dependence*, which means one column is a linear combination of the other columns. This linear dependence can be seen geometrically: when placing the n categories in n-dimensional space, they span an $(n-1)$-dimensional plane only. In Figure 1-5, linear combinations of the three vectors on the left span only the plane containing the triangle, whereas on the right the linear combinations span the full two-dimensional coordinate system.

Multicollinearity causes computational problems for linear and logistic regression, so for those algorithms, we should use $n-1$ dummy variables rather than all n. Even for methods that don't run into this specific computational issue, using fewer independent variables when possible is generally better because this helps reduce the curse of dimensionality—a fundamental topic in data science that we'll visit from a geometric perspective shortly.

On the other hand, for algorithms like k-NN, where distances between data points are crucial, we don't want to drop one of the dummy variables, as that would skew the distances (as we saw in Figure 1-5) and lead to subpar performance. There is a time and a place for both dummy variable methods, and thinking geometrically can help us decide which to use when.

After using dummy variables to convert all categorical variables to numerical variables, we are ready to consider the geometry of the spreadsheet.

The Geometry of Numerical Spreadsheets

We can think of a numerical spreadsheet as describing a collection of points (one for each row) in a *Euclidean vector space*, $\mathbf{R}^d$, which is a geometric space that looks like a flat piece of paper in two dimensions and a solid brick in three dimensions but extends infinitely in all directions and can be any dimension. Here, d is the number of columns, which is also the dimension of the space. Each column in the numerical dataset represents an axis in this space. Concretely, the d-dimensional coordinates of each data point are simply the values in that row.

When $d = 1$, this Euclidean vector space $\mathbf{R}^d$ is a line. When $d = 2$, it is a plane. When $d = 3$, it is the usual three-dimensional space we are used to thinking in. While humans can't really visualize more than three perpendicular axes, higher-dimensional geometry can be analyzed with mathematical and computational tools regardless. It is important to recognize here that just as there are many two-dimensional shapes beyond a flat plane (for instance, the surface of a sphere or of a donut, or even stranger ones like a Möbius strip), there are many three-dimensional geometric spaces beyond the familiar Euclidean space (such as the inside of a sphere in three dimensions or the surface of a sphere in four dimensions). This holds for higher-dimensional spaces as well. Working with structured data has traditionally meant viewing the data from the perspective of $\mathbf{R}^d$ rather than any of these other kinds of geometric spaces.

The Euclidean vector space structure of $\mathbf{R}^d$ is powerful; it allows us to compute all kinds of useful things. We can compute the distance between any pair of data points, which is necessary for a wide range of machine learning algorithms. We can compute the line segment connecting any two points, which is used by the Synthetic Minority Oversampling Technique (SMOTE) to adjust training samples with imbalanced classes. We can compute the mean of each coordinate in the data, which is helpful for *imputing* missing values (that is, filling in missing values with best guesses for the true value based on known data).

However, this nice Euclidean vector space structure is also specific and rigid. Thankfully, we can compute distances between points, shortest paths connecting points, and various forms of interpolation in much more general settings where we don't have global Euclidean coordinates, including *manifolds* (geometric objects like spheres that when zoomed in look like the usual Euclidean space but globally can have much more interesting shape and structure—to come in Chapter 5) and *networks* (relational structures formally introduced in Chapter 2).

As a concrete example, suppose you are working with large-scale geospatial data, such as ZIP code–based crime statistics. How do you cluster data points or make any kind of predictive model? The most straightforward approach is to use latitude and longitude as variables to convey the geospatial aspect of the data. But problems quickly arise because this approach projects the round Earth down to a flat plane in a way that distorts the distances quite significantly. For instance, longitude ranges from −180° to +180°, so two points on opposite sides of the prime meridian could be very close to each other in terms of miles but extremely far from each other in terms of longitudes (see Figure 1-6). It's helpful to have machine learning algorithms that can work on spherical data without the need to map data onto a flat surface.

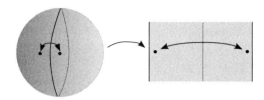

Figure 1-6: Using latitude and longitude (left) as variables for geospatial data distorts distances between data points. Shown here, very near points on opposite sides of the prime meridian are represented by very far points in the longitude-latitude plane (right).

Even when you are working with data that is already structured as a tabular spreadsheet, there might be hidden geometry that is relevant. For example, imagine you have three numerical variables (so that your data lives in $\mathbf{R}^3$) but all your data points live on or near the surface of a sphere in this three-dimensional space. Would you want to consider distances between points as the path lengths along the sphere's surface (which is what is done in spherical geometry) or straight-line distances that cut through the sphere (which is what is done with traditional Euclidean machine learning)? The answer depends on the context and is something that data scientists must decide based on domain knowledge—it is not typically something that an algorithm should decide on its own. For example, if your data points represent locations in a room that an aerial drone could visit, then Euclidean distance is better; if your data points represent airports across the globe that an international airline services, then spherical geometry is better.

One of the main tasks of topological and geometric data science is discovering the geometric objects on which your data points naturally live (like the sphere in the airport example, but perhaps very complex shapes in high dimensions). The other main task is exploiting these geometric objects, which usually involves one or more of the following:

- Applying versions of the usual machine learning algorithms that have been adapted to more general geometric settings
- Applying new geometrically powered algorithms that are based on the shape of data
- Providing meaningful global coordinates to transform your data into a structured spreadsheet in a way that traditional statistical and machine learning tools can be successfully applied

The main goal of this book is to carefully go through all these ideas and show you how to implement them easily and effectively. But first, in the remainder of this introductory chapter, we'll explain some of the geometry involved in a few traditional data science topics (like we did earlier with dummy variables). We'll also hint at the geometry involved in a few different types of "unstructured" data, as a preview of what's to come later in the book.

The Geometry of Supervised Learning

In this section, we'll provide a geometric view of a few standard machine learning topics: classification, regression, overfitting, and the curse of dimensionality.

Classification

Once a dataset has been converted to a numerical spreadsheet (with d columns, let's say), the job of a supervised classifier is to label the predicted class of each new input data point. This can be viewed in terms of *decision boundaries*, which means we carve up the space $\mathbf{R}^d$ into nonoverlapping regions and assign a class to each region, indicating the label that the classifier will predict for all points in the region. (Note that the same class can be assigned to multiple regions.) The type of geometric shapes that are allowed for the regions is determined by the choice of supervised classifier method, while the particular details of these shapes are learned from the data in the training process. This provides an illuminating geometric window into the classification process.

In Figure 1-7, we see decision boundaries for a few standard classification algorithms in a simple binary classification example when $d = 2$.

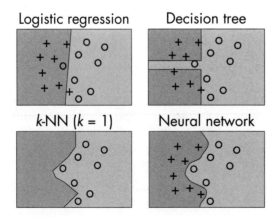

Figure 1-7: The decision boundaries in two dimensions for a few classification algorithms

Logistic regression produces linear decision boundaries. (Though, by adding higher-order terms to the model, you can achieve nonlinear decision boundaries.) Decision trees are built by splitting the independent variables with individual inequalities, which results in decision boundaries made up of horizontal and vertical line segments. In higher dimensions, instead of horizontal and vertical lines, we have planes that are aligned with the coordinate axes. Random forests, which are ensembles of decision trees, still produce decision boundaries of this form, but they tend to involve many more pieces, producing curved-looking shapes that are really made out of lots of small horizontal and vertical segments. The k-NN classifiers produce polygonal decision boundaries since they carve up the space based on which of the finitely many training data points are closest. Neural networks can produce complex, curving decision boundaries; this high level of flexibility is both a blessing and a curse because it can lead to overfitting if you're not careful (we'll discuss overfitting shortly).

Studying the decision boundaries produced by different classifier algorithms can help you better understand how each algorithm works; it can also help you choose which algorithm to use based on how your data looks (for higher-dimensional data, where $d > 2$, you can get two-dimensional snapshots of the data by plotting different pairs of variables). Just remember that there are many choices involved—which variables to use, whether to include higher-order terms, what values to set for the hyperparameters, and so on—and all of these choices influence the types of decision boundaries that are possible. Whenever you encounter a classification algorithm that you aren't familiar with yet, one of the best ways to develop an intuition for it is to plot the decision boundaries it produces and see how they vary as you adjust the hyperparameters.

Regression

Supervised regression can also be viewed geometrically, though it is a little harder to visualize. Rather than carving up space into finitely many regions based on the class predictions, regression algorithms assign a numerical value to each point in the space; when $d = 2$, this can be plotted as a heatmap or a three-dimensional surface. Figure 1-8 shows an example of this, where we first create 10 random points with random dependent variable values (shown in the top plot with the circle size indicating the value), then we 3D scatterplot the predicted values for a dense grid of points, and finally we shade according to height when using k-NN with $k = 3$ (bottom left) and a random forest (bottom right).

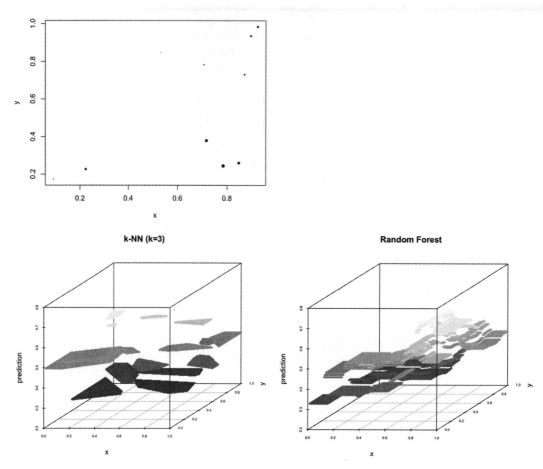

Figure 1-8: The training data with dependent variable values indicated by circle size (top plot), and the three-dimensional prediction surfaces for two nonlinear regression algorithms: 3-NN (bottom left) and random forest (bottom right)

The prediction surface for linear regression (not shown here) is one large, slanted plane, whereas for the two methods illustrated here the surface is a collection of finitely many flat regions for which the prediction value remains constant. Notice that these regions are polygons for *k*-NN and rectangles for the random forest; this will always be the case. Also, for the particular choice of hyperparameters used in this example, the regions here are smaller for the random forest than for *k*-NN. Put another way, the random forest here slices up the data space with surgical precision compared to the *k*-NN algorithm; the latter is more like carving a pumpkin with a butcher's knife. But this will not always be the case—this comparison of coarseness depends on the number of trees used in the random forest and the number of neighbors used in *k*-NN. Importantly, a finer partition for regression is like a more flexible decision boundary for classification: it often looks good on the training data but then generalizes poorly to new data. This brings us to our next topic.

Overfitting

Let's return to the decision boundary plots in Figure 1-7. At first glance, it would seem that the more flexible the boundaries are, the better the algorithm performs. This is true when considering the training data, but what really matters is how well algorithms perform on test data. A well-known issue in predictive analytics is *overfitting*, which is when a predictive algorithm is so flexible that it learns the particular details of the training data and, in doing so, will actually end up performing worse on new unseen data.

In Figure 1-7, the logistic regression algorithm misclassifies the leftmost circle, whereas the decision tree creates a long sliver-shaped region to correctly classify that single point. If circles tend to fall on the left and pluses on the right, then it's quite likely that creating this sliver region will hurt the classification performance overall when it is used on new data—and if so, this is an example of the decision tree overfitting the training data.

In general, as a predictive algorithm's flexibility increases, the training error tends to keep decreasing until it eventually stabilizes, whereas the test error tends to decrease at first and then reaches a minimum and then increases (see Figure 1-9). We want the bottom of the test error curve: that's the best predictive performance we're able to achieve. It occurs when the algorithm is flexible enough to fit the true shape of the data distribution but rigid enough that it doesn't learn spurious details specific to the training data.

Figure 1-9: A plot of training error versus test error as a function of a predictive algorithm's flexibility, illustrating the concept of overfitting

The general behavior illustrated in Figure 1-9 often occurs when varying hyperparameters: the classifier decision boundaries become more flexible as the number of neurons in a neural network increases, as the number of neighbors k in k-NN decreases, as the number of branches in a decision tree increases, and so on.

The Curse of Dimensionality

Sometimes, it helps to think of the x-axis in Figure 1-9 as indicating complexity rather than flexibility. More flexible algorithms tend to be more complex, and vice versa. One of the simplest yet most important measures of the complexity of an algorithm is the number of independent variables it uses. This is also called the *dimensionality* of the data. If the number of independent variables is d, then we can think of the algorithm as inputting points in a d-dimensional Euclidean vector space $\mathbf{R}^d$. For a fixed number of data points, using too few independent variables tends to cause underfitting, whereas using too many tends to cause overfitting. Thus, Figure 1-9 can also be interpreted as showing what happens to a predictive algorithm's error scores as the dimensionality of the data increases without increasing the size of the dataset.

This eventual increase in test error as dimensionality increases is an instance of a general phenomenon known as the *curse of dimensionality*. When dealing with structured data where the number of columns is large relative to the number of rows (a common situation in genomics, among other areas), overfitting is likely for predictive algorithms, and the numerical linear algebra driving many machine learning methods breaks down. This is an enormous problem, and many techniques have been developed to help counteract it—some of which will come up later in this book. For now, let's see how geometry can shed some light on the curse of dimensionality.

One way to understand the curse of dimensionality is to think of Euclidean distances, meaning straight-line distance as the bird flies in however many dimensions exist. Imagine two pairs of points drawn on a square sheet of paper, where the points in one pair are near each other and the points in the other pair are far from each other, as in Figure 1-10.

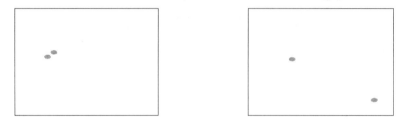

Figure 1-10: A plot of two pairs of points in a two-dimensional space

Let's perturb these points a bit by adding some Gaussian noise; that is, we'll draw four vectors from a bivariate normal distribution and add these to the coordinates of the four points. Doing this moves the points slightly in random directions. Let's do this many times, and each time we'll record the Euclidean distance between the left pair after perturbation and also between the right pair. If the perturbation is large enough, we might occasionally end up with the points on the left farther from each other than the points on the right, but, overall, the Euclidean distances for the left perturbations will be smaller than those for the right perturbations, as we see in the histogram in Figure 1-11.

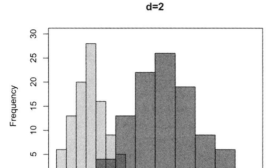

Figure 1-11: A histogram of the Euclidean distances after random small perturbations for the nearby points on the left side of Figure 1-10 (shown in light gray) and the faraway points on the right side of that figure (shown in dark gray)

Next, let's embed our square sheet of paper as a two-dimensional plane inside a higher-dimensional Euclidean space $\mathbf{R}^d$ and then repeat this experiment of perturbing the points and computing Euclidean distances. In higher dimensions, these perturbations take place in more directions. Concretely, you can think of this as padding the x and y coordinates for our points with d-2 zeros and then adding a small amount of noise to each of the d coordinates. Figure 1-12 shows the resulting histograms of Euclidean distances when doing this process for $d = 10$ and $d = 100$.

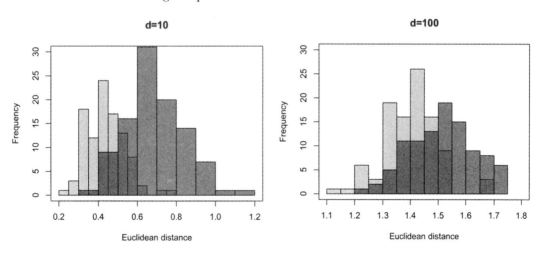

Figure 1-12: Histograms of Euclidean distances as in Figure 1-11, except after embedding in d = 10 dimensions (left) and d = 100 dimensions (right)

We see that as the dimension *d* increases, the two distributions come together and overlap more and more. Consequently, when there is noise involved (as there always is in the real world), adding extra dimensions destroys our ability to discern between the close pair of points and the far pair. Put another way, the signal in your data will become increasingly lost to the noise as the dimensionality of your data increases, unless you are certain that these additional dimensions contain additional signal. This is a pretty remarkable insight that we see revealed here through relatively simple Euclidean geometry!

We can also use perturbations to see why large dimensionality can lead to overfitting. In Figure 1-13, on the left, we see four points in the plane $\mathbf{R}^2$ labeled by two classes in a configuration that is not linearly separable (meaning a logistic regression classifier without higher-order terms won't be able to correctly class all these points).

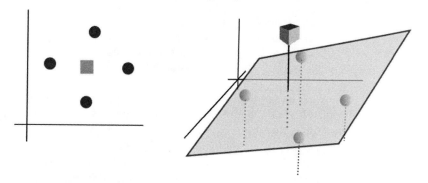

Figure 1-13: Two classes of data points that are not linearly separable in two dimensions (left) but are linearly separable after placing them in three dimensions (right) and perturbing them there

Even if we perturb the points with a small amount of noise, there will be no line separating the two classes. On the right side of this figure, we have embedded these points in $\mathbf{R}^3$ simply by adding a third coordinate to each point that is equal to a constant. (Geometrically, this means we lay the original $\mathbf{R}^2$ down flat above the *xy*-plane in this three-dimensional space.) We then perturb the points a small amount. After this particular perturbation in three dimensions, we see that the two classes do become linearly separable, meaning a logistic regression classifier will be able to achieve 100 percent accuracy here. (In the figure, we have sketched a slanted plane that separates the two classes.)

At first glance, this additional flexibility seems like a good thing (and sometimes it can be!) since it allowed us to increase our training accuracy. But notice that our classifier in three dimensions didn't learn a meaningful way to separate the classes in a way likely to generalize on new, unseen data. It really just learned the vertical noise from one particular perturbation.

In other words, increasing the dimensionality of data tends to increase the likelihood that classifiers will fit noise in the training data, and this is a recipe for overfitting.

There is also a geometric perspective of the computational challenges caused by the curse of dimensionality. Imagine a square with a length of 10 units on both sides, giving an area of 100 units. If we add another axis, we'll get a volume of 1,000 units. Add another, and we'll have a four-dimensional cube with a four-dimensional volume of 10,000 units. This means data becomes more spread out—sparser—as the dimensionality increases. If we take a relatively dense dataset with 100 points in a low-dimensional space and place it in a space with 1,000 dimensions, then there will be a lot of the space that isn't near any of those 100 points. Someone wandering about in that space looking for points may not find one without a lot of effort. If there's a finite time frame for the person to look, they might not find a point within that time frame. Simply put, computations are harder in higher dimensions because there are more coordinates to keep track of and data points are harder to find.

In the following chapters, we'll look at a few different ways to wrangle high-dimensional data through geometry, including ways to reduce the data's dimensionality, create algorithms that model the data geometry explicitly to fit models, and calculate distances in ways that work better than Euclidean distance in high-dimensional spaces. The branch of machine learning algorithms designed to handle high-dimensional data is still growing thanks to subjects like genomics and proteomics, where datasets typically have millions or billions of independent variables. It is said that necessity is the mother of invention, and indeed many machine learning methods have been invented out of the necessity of dealing with high-dimensional real-world datasets.

Unstructured Data

Most of the data that exists today does not naturally live in a spreadsheet format. Examples include text data, network data, image data, and even video or sound clip data. Each of these formats comes with its own geometry and analytic challenges. Let's start exploring some of these types of unstructured data and see how geometry can help us understand and model the data.

Network Data

In the next chapter, you'll get the official definitions related to networks, but you might already have a sense of networks from dealing with social media. Facebook friendships form an undirected network (nodes as Facebook users and edges as friendships among them), and Twitter accounts form a directed network (directional edges because you have both followers and accounts you follow). There is nothing inherently Euclidean or spreadsheet structured about network data. In recent years, deep learning has been extended from the usual Euclidean spreadsheet setting to something much more general called *Riemannian manifolds* (which we'll get to in Chapter 5); the main application of this generalization (called *geometric deep learning*) has been network data, especially for social media analytics.

For instance, Facebook uses geometric deep learning algorithms to automatically detect fake "bot" accounts. In addition to looking at traditional structured data associated with each account such as demographics and number of friends, these detection algorithms use the rich non-Euclidean geometry of each account's network of friends. Intuitively speaking, it's easy to create fake accounts that have realistic-looking interests and numbers of friends, but it's hard to do this in a way such that these accounts' friendship networks are structured similarly to the organic friendship networks formed by real people. Network geometry provides ways of measuring this notion of "similarity."

Geometric deep learning has also been used to detect fake news on Twitter by transforming the detailed propagation patterns of stories through the network into independent variables for a supervised learning algorithm. We won't get to geometric deep learning in this book, but there is still plenty to do and say when it comes to working with network data. For example, we can use geometric properties of a network to extract numerical variables that bring network data back into the familiar territory of structured data.

Image Data

Another form of "unstructured" data that actually has a rich geometric structure is image data. You can think of each pixel in an image as a numerical variable if the image is grayscale or as three variables if it is color (red, green, and blue values). We can then try to use these variables to cluster images with an unsupervised algorithm or classify them with a supervised algorithm. But the problem when doing this is that there is no spatial awareness. A pair of adjacent pixels is treated the same as a pair of pixels on opposite sides of the image. A large branch of deep learning, called *convolutional neural networks* (CNNs), has been developed to bring spatial awareness into the picture. CNNs create new variables from the pixel values by sliding small windows around the image. Success in this realm is largely what brought widespread public acclaim to deep learning, as CNNs smashed all the previous records in image recognition and classification tasks.

Let's consider a simple case of two images that could be included in a larger animal classification dataset used in conservation efforts (see Figure 1-14).

Figure 1-14: An elephant (left) and a lioness (right) at Kruger National Park

The animals are shown in natural environments where leaves, branches, and lighting vary. They have different resolutions. The colors of each animal vary. The extant shapes related to both the animals and the other stuff near the animals differ. Manually deriving meaningful independent variables to classify these images would be difficult. Thankfully, CNNs are built to handle such image data and to automatically create useful independent variables.

The basic idea is to consider each image as a mathematical surface (see Figure 1-15) and then walk across this surface creating a map of its salient features—peaks, valleys, and other relevant geometric occurrences. The next layer in the CNN walks across this map and creates a map of its salient features, which is then fed to the next layer, and so on. In the end, the CNN converts each image to a sequence of maps that hierarchically encode the image's content, with the final layer being the map that is actually used for classification. For these animal images, the first map might identify high-contrast regions in the image. The next map might assemble these regions into outlines of shapes. The following map might indicate which of these shapes are animals. Another layer might locate specific anatomical features within the animals—and these anatomical features could then form the basis for the final species classification.

The precise way the CNN builds these maps is learned internally through the supervised training process: as the algorithm is fed labeled data, connections between neurons in each layer forge, break, and forge again until the final layer is as helpful as possible for the classification task. We'll further explore CNNs and their quantum versions in Chapter 10.

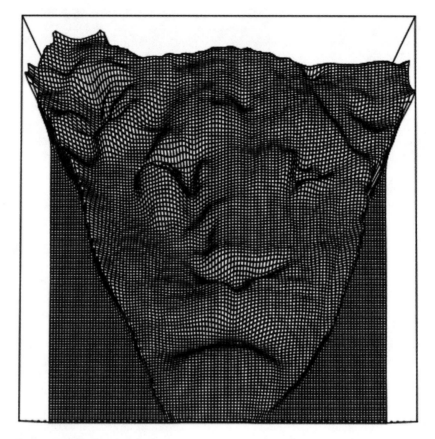

Figure 1-15: The head of the lioness in Figure 1-14, viewed geometrically as a 3D mathematical surface

Using methods from computational geometry to quantify peaks and valleys has applications beyond image recognition and classification. A scientist might want to understand the dynamic process or structure of a scientific phenomenon, such as the flow of water or light on an object. The peaks, valleys, and contours of the object impact how light will scatter when it hits the object, and they'll also determine how liquids would flow down the object. We'll cover how to mine data for relevant peaks, valleys, and contours later in this book.

Text Data

Another form of "unstructured" data that has risen to prominence in recent years is text data. Here, the structure that comes with the data is not spatial like it is for images; it's linguistic. State-of-the-art text processing (for instance, used by Google to process search phrases or by Facebook and Twitter to detect posts that violate platform policies) harnesses deep learning to create something called *vector embeddings*, which translate text into

$\mathbf{R}^d$, where each word or sentence is represented as a point in a Euclidean vector space. The coordinates of each word or sentence are learned from data by reading vast amounts of text, and the deep learning algorithm chooses them in a way that in essence translates linguistic meaning into geometric meaning. We'll explore deep learning text embeddings in Chapter 9.

For example, we might want to visualize different sets of variables concerning text documents. Because the variables form a high-dimensional space, we can't plot them in a way that humans can visualize. In later chapters, we'll learn about geometric ways to map high-dimensional data into lower-dimensional spaces such that the data can be visualized easily in a plot. We can decorate these plots with colors or different shapes based on the document type or other relevant document properties. If similar documents cluster together in these plots, it's likely that some of the variables involved will help us distinguish between documents. New documents with unknown properties but measured values for these variables can then be grouped by a classification algorithm. We'll explore this further in Chapter 9.

Summary

This chapter provided a brief review of the main concepts of traditional machine learning, but it put a geometric spin on these concepts that will likely be new for most readers. Woven into this review was a discussion of what it means for data to be structured. The main takeaway is that essentially all data has meaningful structure, but this structure is often of a geometric nature, and geometric tools are needed to put the data into a more traditional spreadsheet format. This is a theme we'll develop in much more depth throughout the book. This chapter also hinted at some of the important geometry hidden in spreadsheet data. One of the main goals of the book is to show how to use this hidden geometry to improve the performance of machine learning algorithms.

We'll start in Chapters 2 and 3 by diving into algorithms designed for analyzing network data, including social and geographic networks. This includes local and global metrics to understand network structure and the role of individuals in the network, clustering methods developed for use on network data, link prediction algorithms to suggest new edges in a network, and tools for understanding how processes or epidemics spread through networks.

2

THE GEOMETRIC STRUCTURE OF NETWORKS

Networks, which we briefly encountered in Chapter 1, play an increasingly large role in our lives; accordingly, network data plays an increasingly large role in data science. How do we measure the influence of a user on social media? Or judge the robustness of a computer network against hackers? Or identify people who bridge different social groups? These are all questions about the geometric structure of networks, and they are all examples of concepts we will explore in this chapter.

We'll begin this chapter with a brief motivational section arguing why network data is an important topic in modern data science that greatly benefits from geometric reasoning. We'll then introduce the basic concepts and definitions in network theory, which is rooted in the mathematical language of graph theory. The bulk of the chapter will be a guided tour of various quantities associated to networks and their vertices. We'll conclude with a quick look at a few different types of random networks that have been studied extensively in the literature and are easy to generate in the language R.

Introduction to Network Science

Network science, which studies entities through their relationships with each other, is an important interdisciplinary subject that has gained momentum in data science since the rise of social networks. As we saw in Chapter 1, network data is "unstructured" in the usual sense of the term, but it is highly structured in other ways. Mathematically, networks are rooted in a subject called *graph theory*, which distills entities and relationships down to their abstract essence. Networks are geometrically rich, but their geometry is different from the usual Euclidean geometry of angles and straight lines. To extract insights from networks, we must leave the Euclidean world of spreadsheets and embrace a more exotic geometry where spheres look nothing like the round objects we're used to seeing.

Let's consider a social media example. Users on a platform such as Facebook form a network, and distance can be defined by the smallest number of friendships connecting two individuals. This notion of distance underpins several ways of measuring a user's centrality in the network, which we can use to quantify "influencers" in various ways. For instance, we could count the number of friends directly connected to a Facebook user's account, or we could count how many friends of friends exist. Someone may not have many immediate connections but may have a large number of friends who are well connected. In other words, a user may have a large sphere of influence even if their immediate circle of friends is small. The geometry of networks also allows us to detect outliers by measuring how improbable a user's network is compared to others; this helps social media platforms automatically detect inauthentic accounts.

A fascinating application of network-based outlier detection arose in a recent state supreme court case on gerrymandering. A conspicuous map of congressional districts in Pennsylvania prior to 2018 was widely believed to be the result of a partisan gerrymander favoring Republicans, but the court was not convinced until a team of experts brought in network science. They showed that one can view a districting map as a network, and by computing a random walk in the space of map networks to statistically probe the geometry of these networks, they found that the Pennsylvania map was several standard deviations away from the mean. The new map under consideration at the time was right in the middle of the bell curve of maps. They concluded that the old map was highly unlikely to have arisen organically; it was so unusual that it must have been deliberately engineered by the mapmakers (who were beholden to current Republicans in office) to increase the number of seats Republicans would hold. In contrast, the new map was much more typical (and a fair representation). The court was convinced by this network analysis and threw out the old map in 2018 to adopt the new one.

Even in scenarios that do not outwardly appear to be about networks, there is sometimes a way of viewing data as network data—and doing so avails one of a rich set of tools that draw from the remarkable geometry of networks.

This chapter focuses on the geometry of networks and how it can be used to define a wide collection of metrics that quantify various aspects of the shape and structure of a network. In the next chapter, we'll apply these metrics to explore machine learning and other forms of data analysis on network data. Let's officially begin our journey.

The Basics of Network Theory

A network conveys pairwise relationships between entities, which can be individuals, objects, items, and so on. The entities are represented by *vertices* (also called *nodes*), while a relationship between a pair of entities is represented by an *edge* connecting the corresponding pair of vertices. Here are a few examples of entities and relationships that have been studied as networks: phone calls and text messages between people, proteins that interact in a biological pathway, towns connected by roads, websites connected by page links, parts of the brain activated during the same task, connections between words or parts of speech in a language—and that's just scratching the surface of popular examples.

When plotting networks, the vertices are depicted as dots that are sometimes decorated with symbols so we know which vertex is which; the edges are usually straight-line segments, although sometimes it is helpful (for instance, to avoid artificial edge crossings) to draw some edges instead as curved arcs. All that matters is the collection of vertices and the edges between them—not how long the edges are in the plot, the angles between them, or the locations of the vertices. In fact, as shown in Figure 2-1, the same network can be drawn in rather different ways.

These have the same vertices and the same edges connecting the vertices, but the plotting of these relationships varies. However we visualize the relationships, their underlying structure does not change.

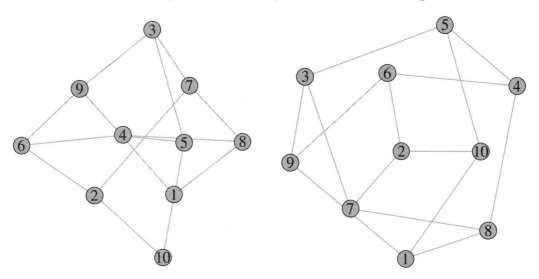

Figure 2-1: Two different plots of the same network. These have the same vertices and the same edges connecting vertices, but the plotting of these relationships varies. However we visualize the relationships, their underlying structure does not change.

Directed Networks

A network is *directed* if the edges represent one-way relationships (meaning they point from one vertex, called the *source*, to the other vertex, called the *target*); otherwise, the edges represent mutual relationships, and the network is *undirected*. For instance, for phone conversations, we could create an undirected network with an edge between any two people who have talked to each other, or we could create a directed network with edges representing outgoing calls from the source person to the target person.

Concretely, suppose we have three work colleagues: Sadako, Pedro, and Georg. Sadako is the project lead and, as such, makes calls to Pedro and Georg. Pedro calls Georg regarding a question on the engineering side of the project, but Georg does not call Pedro or Sadako. Pedro does not call Sadako, as he has already received instructions when she calls him. Figure 2-2(a) shows the undirected network for these colleagues, while Figure 2-2(b) shows the directed network.

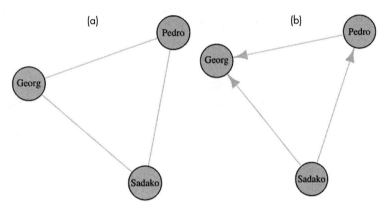

Figure 2-2: A phone network among three colleagues that is (a) undirected and (b) directed

Other examples of directed networks include travel routes, machine and social processes, needle sharing to trace epidemics, and biological process models. Mathematically speaking, the biggest difference between Facebook and Twitter is that Facebook is an undirected network (friendships are mutual), while Twitter is a directed network (users have both followers and accounts followed).

Networks in R

There are two main ways of representing a network in a computer. An *edge list* simply lists, in an arbitrary order, all the edges in the network by naming the two vertices of each edge (if the network is directed, then the first vertex is considered the source and the second vertex the target). A useful package for working with network data in R is the igraph library (which also exists in Python with similar commands and syntax). To create the directed network in Figure 2-2(b) from an edge list, you can use the code in Listing 2-1.

```
library(igraph)
edges<-rbind(c("Sadako","Pedro"),c("Sadako","Georg"),c("Pedro","Georg"))
g_colleagues<-graph.edge(edges,directed=T)
```

Listing 2-1: A script that generates the directed network in Figure 2-2(b) from an edge list

The other representation of a network is a spreadsheet. If a network has n vertices, its *adjacency matrix* is the $n \times n$ matrix whose rows and columns are indexed by the vertices where 0 in the (i, j) entry means there is no edge from vertex i to vertex j, while 1 in this entry means there is such an edge. If the network is undirected, then the adjacency matrix is symmetric (equal to its own transpose, meaning it is unchanged when swapping its rows with its columns) because, in that case, having an edge from vertex i to vertex j is the same as having an edge from vertex j to vertex i. Listing 2-2 provides the R code to construct the preceding directed network from an adjacency matrix rather than an edge list.

```
library(igraph)
adjmat<-matrix(c(0,0,0,1,0,0,1,1,0),nrow=3)
rownames(adjmat)<-c("Sadako","Pedro","Georg")
colnames(adjmat)<-c("Sadako","Pedro","Georg")
g_colleagues<-graph_from_adjacency_matrix(adjmat,mode="directed",weighted=T)
```

Listing 2-2: A script that generates the directed network in Figure 2-2(b) from an adjacency matrix

A major advantage of the adjacency matrix approach is that it puts networks into the framework of linear algebra, where many powerful tools are available. For instance, we can use the spectral theory (eigenvalues and eigenvectors) of the adjacency matrix to measure centrality in a network. Adjacency matrices also readily generalize to *weighted networks*; by allowing arbitrary numbers as the entries (not just the binary 0 and 1 entries discussed so far), we can assign a real-number weight to each edge that conveys the strength of the relationship represented by that edge. In practice, this can be used to represent many different important notions: volume of calls between colleagues in a phone network, distances between locations in a transportation network, and so on. You can think of an unweighted network as a weighted network in which all the edges have a weight equal to 1. In fact, this is why in Listing 2-2 we set the `weighted` parameter to T (true) even though we wanted to create an unweighted network—otherwise, igraph interprets the adjacency matrix differently.

Paths and Distance in a Network

Two vertices are said to be *neighbors* if they are connected by an edge. Two edges in a network are *adjacent* if they have a vertex in common. A *path* in a network is a sequence of adjacent edges. If the network is directed, then all the edges in a path must be compatibly oriented. For example, in Figure 2-2(b), we have a two-edge path from Sadako to Pedro to Georg and a one-edge path from Pedro to Georg, but going from Pedro to Sadako to Georg is not a path since the edges are not oriented this way.

The *length* of a path has two different meanings, depending on whether the network is weighted or unweighted. In an unweighted network, the length of a path is the number of edges in the path; in a weighted network, the length is the sum of the weights of the edges in the path. The *network distance* between two vertices is the length of the minimal-length path connecting them, assuming there is a path from one to the other (if there is no path, the network distance is undefined). People—us included—usually just say "shortest" path instead of "minimal-length" path, because it sounds more natural. This is somewhat misleading for weighted networks, as the minimal-length path is actually the path with the lowest weight, which may not be the shortest path in terms of the number of edges in the path. However, that's a minor imperfection in terminology we can all live with if we're careful.

As an example, let's consider the weighted undirected network in Figure 2-3, representing four towns and the length (in miles) of the roads between them.

Plot of Connected Towns by Road Miles

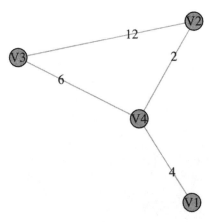

Figure 2-3: A plot of town connectivity and miles of road between connected towns

The shortest path between V2 and V3 is the 8-mile path passing through V4, rather than the 12-mile road directly connecting these two towns. The network distance between V2 and V3 is therefore 8 miles rather than 12. Note that the physical lengths of the edges and paths in this plot do not in any way represent the lengths as defined by the weights; in other words, Euclidean distance in a network plot has nothing to do with network distance! The placement of the vertices in a network plotted by igraph is chosen for aesthetic reasons (to minimize the number of edge crossings, for instance) rather than to reflect edge weights. This can be tricky to get used to at first but over time will start to feel natural. Listing 2-3 provides the R code to create and plot the network in this example.

```
#create symmetric matrix of town connections and miles between each town
towns<-matrix(c(0,0,0,4,0,0,12,2,0,12,0,6,4,2,6,0),nrow=4)

#label the rows and columns so the towns have names
rownames(towns)<-c("V1","V2","V3","V4")
colnames(towns)<-c("V1","V2","V3","V4")

#create a weighted undirected network from this adjacency matrix
library(igraph)
g_towns<-graph_from_adjacency_matrix(towns,mode="undirected",weighted=T)

#plot town graph with edges labeled by weights
plot(g_towns,edge.label=E(g_towns)$weight,main="Plot of Connected Towns by Road Miles",
vertex.color=2,vertex.size=20)
```

Listing 2-3: *A script that generates the network of connected towns from Figure 2-3 and plots it*

Network distances can play a major role in real-world problems. During the 2018 Ebola outbreak in the eastern Democratic Republic of the Congo, limited physical routes between towns impacted epidemic spread, supply-chain logistics to move medical supplies and personnel, and population mixing between impacted towns and towns that didn't have active cases at the time.

A word of caution about weighted networks: we tend to think of the vertices that are near each other (defined by network distance) as being the most closely related or the most strongly connected. This means that the smaller the weight of an edge, the stronger the relationship it represents. For this reason, you'll often want to use reciprocals when setting the weights in a network. For instance, in a weighted phone network, rather than setting the weight between person A and person B to be the number of calls c_{AB} between them, you should set it to be the reciprocal $1/c_{AB}$ so that people who talk to each other more frequently will be closer to (rather than farther from) each other in the weighted network. In Chapter 3, we'll revisit the town network from Figure 2-3, and we'll reciprocate the weights so that they represent proximity rather than distance.

To start exploring the geometry established by network distance, we'll turn now to some notions of centrality (which can also be thought of as importance or influence) in a network. Since we will rely on igraph throughout this chapter, we will no longer include library(igraph) in the code snippets; therefore, be sure to load this library before running any of the following code samples.

Network Centrality Metrics

Measuring the *centrality*, or importance, of each vertex in a network allows you to manually analyze the structure and functional behavior of networks. It is also frequently used for feature extraction, as centrality metrics provide numerical features that can be fed into traditional supervised and unsupervised machine learning algorithms, as we'll see in Chapter 3. In other words, centrality metrics provide a way to equip network data with more conventional spreadsheet structure. There are many methods for quantifying vertex centrality, almost all of which involve notions from the preceding section—paths, network distance, and the adjacency matrix.

Centrality has many real-world applications. In epidemiology, individuals who are more central in contact tracing networks tend to spread disease to more people than less central individuals. Indeed, central individuals are typically connected to a wider range of people, and additionally central individuals often connect groups of people who otherwise would be unlikely to meet—thereby providing bridges between these groups across which disease can spread. In social media, central individuals are the "influencers" whose opinions are heard by many and spread widely. In marketing, central individuals can act as lucrative vectors for advertising campaigns. In criminology, identifying central individuals in drug distribution or organized crime networks allows law enforcement agencies to target their actions most effectively. In scientific research, centrality in citation networks helps reveal high-impact publications.

There are more centrality metrics than can be covered in this chapter, so we'll focus on a handful of the popular ones that are implemented in the igraph library.

The Degree of a Vertex

The most basic measure of centrality of a vertex is its *degree*, which by definition is the number of edges attached to the vertex. In a directed network, this can be broken down into two pieces: the *in-degree* counts the number of edges with this vertex as the target vertex, and the *out-degree* counts the number of edges with this vertex as the source vertex. In a Facebook network, your degree is your number of friends; in Twitter, your in-degree is your number of followers, and your out-degree is the number of accounts you follow. Going back to Figure 2-3, towns V2 and V3 each have degree 2, while V1 has degree 1 and V4 has degree 3.

In a weighted network, you can also measure the *strength* (also called the *weighted degree*) of a vertex, which is the sum of the weights of the edges attached to the vertex. In Figure 2-3, towns V1, V2, V3, and V4 have strengths 4, 14, 18, and 12, respectively. Since an unweighted network can be viewed as a weighted network in which all edges have weight 1, the vertex degree really is just a special case of strength.

Despite its simplicity, the degree of a vertex is a valuable metric. It is the one numerical quantity most social media platforms publicly list for each account, and it is typically used to determine how much an influencer is paid in a marketing campaign to promote a product. However, it has impactful limitations. The degree of a vertex measures the size of that vertex's immediate network of connections, but it does not capture the structure of this network nor does it look beyond these immediate connections. To illustrate why this matters, consider a Twitter user with 10,000 followers, each of whom has only a small number of followers; now compare this with a Twitter user who has only a few hundred followers, but some of these followers are highly influential users. While the first user has a higher degree, which user's tweets are more likely to receive more retweets in the end? This example hints at how networks involve both breadth and depth, and degree is more focused on the former.

For another example of what degree does and does not capture, imagine two individuals in a contact tracing network. Suppose one user has more contacts (and hence a higher degree), but those contacts are mostly all connected to each other anyway, whereas the other individual has fewer contacts, but the contacts are spread across multiple communities that otherwise have almost no interaction. (We'll see a concrete example of this situation later in this chapter.) Which individual poses a greater epidemiological risk during an outbreak?

These two limitations of degree—ignoring network depth and ignoring community-bridging properties—are two of the main reasons that we must push deeper into the geometry of networks. So, let's take a brief march through the zoo of vertex centrality measures that go beyond degree.

The Closeness of a Vertex

How does one determine whether a vertex is located near the center of a network or whether it's near the periphery? The *closeness* of a vertex, defined as the reciprocal of the sum of network distances between this vertex and each other vertex in the network, is designed to capture this distinction. A vertex that is near the center of the network will have a relatively modest distance to the other vertices in the network, but a more peripheral vertex will have a modest distance to some vertices but a large distance to the vertices on the "opposite" side of the network. This means the sum of distances for a central vertex is smaller than the sum of distances for a peripheral vertex; reciprocating this sum flips this around so that the closeness score is greater for a central vertex than for a peripheral vertex.

In Figure 2-3, you might intuitively guess that V4 is the most central, V2 and V3 are moderately central, and V1 is the most peripheral. Indeed, if we ignore the edge weights and compute the closeness scores as an unweighted network, we get 0.20, 0.25, 0.25, and 0.33 for V1, V2, V3, and V4, respectively. For instance, the closeness of V2 is $1/(1 + 1 + 2)$. When using the edge weights, we get closeness scores of 0.05, 0.06, 0.04, and 0.08, so V4 is still the most central, but now V3 is the most peripheral; this is because the edge weights make V3 have a relatively large network distance from the other vertices.

The Betweenness of a Vertex

The *betweenness* of a vertex measures centrality by computing how many paths in the network pass through the vertex; more precisely, it is the sum over all pairs of other vertices in the network of the fraction of shortest paths between the pair of vertices that pass through the vertex in question. When considering Figure 2-3 as an unweighted network, the betweenness of V1 is 0 because no shortest paths between the remaining vertices pass through V1. The same is true of V2 and V3. The betweenness of V4, however, is 2 = 1 + 1 + 0, as there is one shortest path between V1 and V2 that passes through V4, there is one shortest path between V1 and V3 that passes through V4, and the shortest path between V2 and V3 does not pass through V4.

In Figure 2-4, each vertex has a betweenness score of 0.5. For instance, for V1 the one-edge path between V2 and V4 does not pass through V1, the one-edge path between V3 and V4 does not pass through V1, and between V2 and V3 there are two shortest paths, one of which passes through V1, combining to give a betweenness score of 0 + 0 + 1/2.

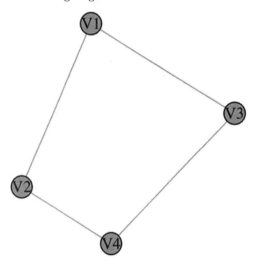

Figure 2-4: A network with four vertices connecting as a square

If you think of a network as encoding the ways that something (materials, information, disease, traffic, and so on) can travel between entities, then shortest paths are the most efficient travel routes. Betweenness centrality measures how disruptive removing a vertex would be to these routes. For example, in a shipping distribution network, the betweenness of a vertex indicates the number of shipping routes that would be impacted if a particular distribution center is shut down (not counting routes originating or terminating at that distribution center), weighted by the extent of this impact. In Figure 2-3 (viewed as an unweighted network), removing V1 does not affect the shipping routes among the other cities (betweenness score of 0), but removing V4 completely takes away two shipping routes (betweenness score of 2). In Figure 2-4, removing V1 impacts transportation between V2 and V3, but it takes out only half of the shipping routes between them (betweenness score of 0.5). Betweenness centrality's use in estimating disruptive potential has found many practical applications; for instance, it identifies important servers to protect in a computer network (or which ones would be most effective to target in an adversarial attack!).

While closeness and betweenness measure rather different aspects of the centrality of a vertex, they are both based on shortest paths in the network. Shortest paths between points have a special name in geometry, *geodesics*, and we'll return to them in much more generality in Chapters 4, 5, and 6.

All the remaining centrality measures that we'll cover in this section are based on the adjacency matrix, and they can be thought of as variants of the famous PageRank algorithm that Google initially used (and still uses in some capacity) to rank search results by estimating importance in the directed network whose vertices are websites and whose edges are links. The math involved in these next centrality measures is heavier, but we'll try to highlight the main ideas and big picture. Practically speaking, it is far more important to develop an intuition for what these centrality scores convey than to understand the details of their computation.

Eigenvector Centrality

Adjacency matrix-based centrality measures aim to capture the basic idea that the importance of a vertex is determined by the importance of the vertices to which it is connected. This is where the notion of *network depth* alluded to earlier comes in. On social media, for instance, it isn't just your number of followers that matters, it is the number of followers your followers have, the number of followers they have, and so on.

Let's make this idea precise. Suppose we want to assign an importance score x_i to each vertex i. A simple model is to assume that the importance of each vertex is proportional to a weighted sum of the importances of the neighboring vertices:

$$x_i = c\sum a_{ij} x_j$$

where the sum is over vertices j that are neighbors to vertex i, a_{ij} is the weight of the edge between vertex i and vertex j, and c is a constant of proportionality that we'll assume is independent of i. This formula can be expressed as the matrix equation:

$$\vec{x} = cA\vec{x}$$

where A is the adjacency matrix of the network. You might recall from linear algebra that this equation stipulates that the vector of vertex importances $\vec{x}$ is an *eigenvector* for the adjacency matrix, and $1/c$ is the corresponding *eigenvalue*. When viewing a matrix as a linear transformation, the eigenvectors are the directions that get stretched but not rotated, and the eigenvalues measure how much these directions are stretched.

Matrices typically have many different eigenvectors and eigenvalues, and the eigenvalues need not be real numbers, so at first glance there could be many different solutions to the preceding vertex assignment problem—some of which may involve complex values. Fortunately, however, if we disallow negative edge weights and importance scores, then something called the *Perron–Frobenius theorem* guarantees that we can find a unique real-valued solution to this vertex assignment problem. In other words, with this non-negativity condition there is always exactly one way of assigning importance scores to all the vertices in the network that satisfy all of our linear neighbor influence conditions. Well, almost: to get uniqueness, we need to normalize the importance scores (for instance, by scaling so that the largest importance score of any vertex in the network is 1). Doing this yields the *eigenvector centrality* scores for vertices.

In Figure 2-3, the eigenvector centrality scores (rounded to the nearest one-hundredth) for V1, V2, V3, and V4, when viewed as an unweighted network, are 0.46, 0.85, 0.85, and 1, respectively. This matches our intuition that V4 is the most central and V1 is the least central. When computed as a weighted network, using the edge weights shown in that figure, the scores are instead 0.16, 0.91, 1, and 0.59. So, in the weighted network, V1 is still the least central, but the most central is V3.

The main point of eigenvector centrality is that it is higher for vertices that neighbor vertices with higher eigenvector centrality scores. In social network terms, this means your importance is determined by the importance of your friends.

We can interpret eigenvector centrality more precisely by thinking in terms of random walks. Imagine you start at a random vertex and repeatedly take steps by choosing one of the edges attached to your current vertex at random. If the network is unweighted, then you choose these edges with equal probability, whereas if the network is weighted, then the probabilities for selecting the edges are proportional to the edge weights. The eigenvector centrality scores are proportional to the fraction of time you spend at each vertex when doing these random walks. For example, if the network represents the World Wide Web (where vertices are websites and edges are links), then eigenvector centrality conveys the amount of traffic each site gets if users just randomly click links.

One drawback with eigenvector centrality is that random walkers can never travel between two different parts of a network if there are no paths connecting the two parts. In other words, each random walker is constrained to the "island" (or, in more mathematical terms, *connected component*) it starts on. This is especially problematic for directed networks because edges can be traversed in only one direction: if the random walker reaches a dead end, it will remain stuck there forever. One way to help get our random walkers past islands and dead ends is to allow them to randomly jump to different locations in the network. That's the main idea behind the next measure we'll look at.

PageRank Centrality

Google's *PageRank centrality* is only a small modification of eigenvector centrality; it replaces the adjacency matrix with a scaled version before computing the eigenvector. The PageRank scores for V1, V2, V3, and V4 in

Figure 2-3, computed as an unweighted network, are 0.13, 0.25, 0.25, and 0.37, respectively.

The best way to understand PageRank is in terms of random walk probabilities. We play the same game as we did with eigenvector centrality, except now at each step there are two possibilities: the random walker either walks across one of the attached edges, as before, or jumps directly to a random vertex in the network. In the World Wide Web example, this means either people can click links or they can directly type URLs in their web browsers. This ability to jump helps the random walkers explore greater extents of the network, which helps boost the information captured by this centrality score compared to eigenvector centrality.

Katz Centrality

Another useful centrality measure that is also computed from the spectral theory of the network's adjacency matrix is *Katz centrality* (sometimes called *Bonacich centrality* or *alpha centrality*), which is essentially a "higher-order" extension of the notion of degree. In an unweighted network, Katz centrality is a weighted sum of the number of vertices that can be reached by a path of length 1 (this coincides with the degree), the number of vertices that can be reached by a path of length 2, the number that can be reached by a path of length 3, and so on. The weights in this weighted sum are determined by the length of the path and by a user-specified "attenuation" parameter (called *alpha*) that ranges from 0 to 1. More precisely, paths of length d are given weight α^d. This means that the farther away a vertex is, the less it contributes to the weighted sum.

For example, in social networks or social media, the biggest contributor to your Katz centrality is your number of friends, the next biggest contributors are the numbers of friends these friends have, and so on. In short, Katz centrality is a more sophisticated version of degree that looks deeper into your network's overall connectivity. It turns out that eigenvector centrality is a limit of Katz centrality as the attenuation parameter approaches a certain value. Be careful when choosing the parameter alpha, though; you should always choose it to be less than the reciprocal of something called the *spectral radius* of the network (which we'll cover in the "Global Network Metrics" section); igraph won't give you an error or even a warning message if you don't do this—you'll just get values that don't make sense, such as negative scores for some vertices.

Hub and Authority

Another way to generalize eigenvector centrality is to provide each vertex with two separate importance scores rather than just one. The most popular incarnation of this is called *authority and hubness*. Conceptually, *authority centrality* measures how much knowledge of the network is stored within a vertex, while *hub centrality* measures how well a vertex knows where to find this knowledge (quick access to the information stored in nearby vertices). These two measures are interrelated: strong hubs tend to connect to strong authorities. Hubs usually have a high rate of connections among the vertices involved that allow for rapid information sharing.

Measuring Centrality in an Example Social Network

To better understand all these vertex centrality measures, it helps to consider a more interesting example of a network than what we've considered so far. The script in Listing 2-4 loads and then plots a network from a file that represents a social network of one of this book's authors (Farrelly). You can find the dataset, along with the other files for the book, on the book's web page: *https://nostarch.com/shapeofdata*.

```
#load data including no header
mydata<-as.matrix(read.csv("SocialNetwork.csv",header=F))

#convert data to graph
g_social<-graph_from_adjacency_matrix(mydata,mode="undirected",weighted=T)

#plot graph
plot(g_social,main="Farrelly's Social Network",vertex.size=15,vertex.label.cex=0.5,
vertex.color=2)
```

Listing 2-4: A script that loads a social network dataset, converts it to a graph object, and then plots it

Figure 2-5 shows the resulting network plot (which will look a little different each time you run this script).

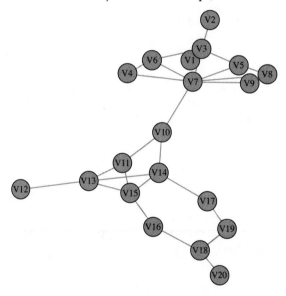

Figure 2-5: A plot of Farrelly's social network dataset, including connections within her medical school service groups (top cluster) and her veterans organization groups (bottom cluster), with vertex 7 (Farrelly herself) serving as a bridge

One of the best ways to develop an intuition for centrality measures is to plot them as the vertex size parameter. (More generally, using vertex size to illustrate any numerical property associated to the vertices of a network is an excellent visualization technique—it is a network version of bubble charts.) Since the range of the different centrality measures varies, in what follows we rescale each one so that the maximum value is 20, as this seems to give fairly readable plots. Listing 2-5 provides the R code we use to plot each one.

```
#betweenness
plot(g_social,vertex.size=20*betweenness(g_social)/max(betweenness(g_social)),
vertex.label.cex=0.8,vertex.color=2)

#closeness
plot(g_social,vertex.size=20*closeness(g_social)/max(closeness(g_social)),
vertex.label.cex=0.8,vertex.color=2)

#eigenvector centrality
plot(g_social,vertex.size=20*eigen_centrality(g_social)$vector,vertex.label.cex=0.8,
vertex.color=2)

#PageRank centrality
plot(g_social,vertex.size=20*page_rank(g_social)$vector/max(page_rank(g_social)$vector),
vertex.label.cex=0.8,vertex.color=2)

#Katz centrality (with alpha parameter set to 0.2)
plot(g_social,vertex.size=20*alpha_centrality(g_social,
alpha=0.2)/max(alpha_centrality(g_social,alpha=0.2)),vertex.label.cex=0.8,vertex.color=2)
```

Listing 2-5: A script that creates bubble chart network plots for each centrality measure on Farrelly's social network dataset

Let's start with closeness and betweenness, illustrated in Figure 2-6.

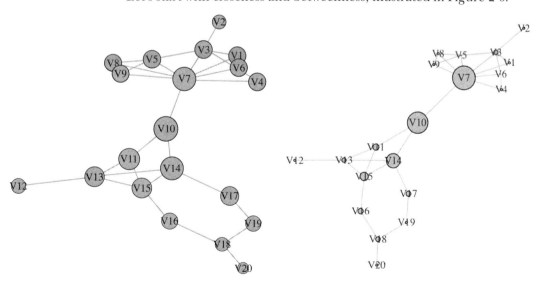

Figure 2-6: Bubble charts of closeness centrality (left) and betweenness centrality (right) on Farrelly's social network data

The closeness scores here appear to capture what we were hoping they would. Vertices that are located more centrally in the network are larger, and vertices that are more peripheral are smaller. Note that this notion of centrality is based on the global geometry of the network. For instance, V10 has the highest closeness centrality score because it is central to the network overall even though it is peripheral to each of the two main clusters (the medical school friends and the veteran friends).

Betweenness centrality (which, as you recall, aims to capture locations that when removed would be maximally disruptive to flow across the network) provides V7 and V10 with much larger scores than all the other vertices—and these are the two vertices that bridge the two main clusters in this network. There is also a more modest but still relatively large betweenness score for V14, indicating that many (but not all) shortest paths between the two main clusters pass through this vertex.

Next, let's take a look at the eigenvector centrality and PageRank centrality scores shown in Figure 2-7.

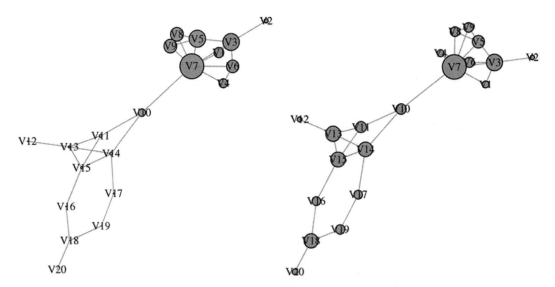

Figure 2-7: Bubble charts of eigenvector centrality (left) and PageRank centrality (right) on Farrelly's social network data

For both of these measures, V7 (representing Farrelly herself) has the highest value, and the values range across the medical school cluster in a way that—at least intuitively—does appear to capture a notion of centrality or importance. Perhaps the most striking thing about this figure is the contrast between these two closely related eigenvector-based measures. For eigenvector centrality, the values in the medical school cluster completely dwarf those of the veterans organization cluster, whereas PageRank centrality seems better able to reflect centrality and importance within each of the two clusters separately. The reason for this discrepancy is that random walks in which one simply chooses neighboring vertices to traverse with equal probability tend to meander around the highly interconnected medical school cluster and have a low probability of taking the one edge (from V7 to V10) that leads out of that cluster. With PageRank centrality, the random walks include a fixed probability of jumping anywhere in the network, which helps them get to the veterans group cluster.

In Figure 2-8 we have Katz centrality, which as you recall is a higher-order version of degree that takes into account not just the number of neighbors but the number of neighbors of neighbors, and so on. We compute this for two different values of the attenuation parameter alpha by specifying the decay rate for the influence coming from higher-order connections.

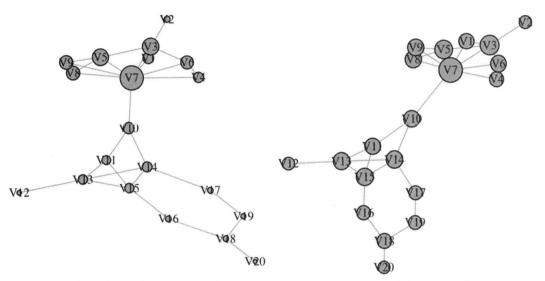

Figure 2-8: Bubble charts of Katz centrality for attenuation parameter alpha = 0.2 (left) and alpha = 0.1 (right) on Farrelly's social network data

It turns out the spectral radius for this network is about 4, so we need to choose values of alpha that are less than 0.25 (here, 0.2 and 0.1). We see here that both choices identified Farrelly's V7 as the most central vertex, but for the higher of these two alpha parameters (left plot), the scores remain relatively high near V7 before dissipating rather quickly. With the smaller alpha (right plot), the dissipation is less sharp, and the Katz centrality scores are more evenly spread out across the entirety of the network. The plot on the left is more likely to capture what we typically think of as centrality or importance, and it is usually recommended to choose an alpha very close to the reciprocal of the spectral radius (and definitely not the default value of 1 in igraph).

Finally, let's analyze authority and hubness. In an undirected network, these two measures coincide, so let's first transform our network into a directed network. Since we don't have a natural direction to the friendships in this social network, we'll do this artificially as follows. For each edge in this network, say between vertex i and vertex j, with probability one out of three, we'll convert it to a one-way edge from i to j, with probability one out of three, we'll convert it to a one-way edge from j to i, and with probability one out of three we'll leave it as a two-way edge. Listing 2-6 shows the code to do this and then plot the resulting hub and authority scores.

```
#randomly remove some entries in the adjacency matrix
mydata_directed<-mydata
for (i in 1:20){
  for (j in i:20){
     rand<-runif(1)
     if(rand < 0.33){mydata_directed[i,j]<-0}
     if(rand >= 0.33 & rand < 0.66){mydata_directed[j,i]<-0}
  }
}

#use this modified adjacency matrix to create a directed network
g_directed<-graph_from_adjacency_matrix(mydata_directed,mode="directed",weighted=T)

#plot the hub centrality and authority centrality for this directed network
plot(g_directed,vertex.size=20*hub_score(g_directed)$vector,vertex.label.cex=0.8,
vertex.color=2,edge.arrow.size=0.4)
plot(g_directed,vertex.size=20*authority_score(g_directed)$vector,vertex.label.cex=0.8,
vertex.color=2,edge.arrow.size=0.4)
```

Listing 2-6: A script that turns Farrelly's social network into a directed network and then plots the resulting hub and authority scores

Figure 2-9 shows the resulting plots.

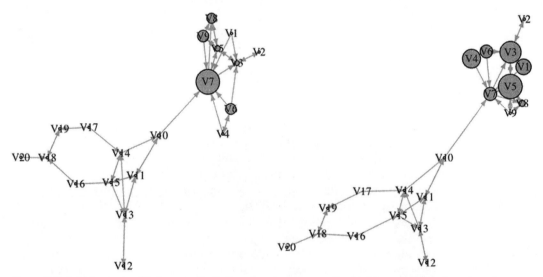

Figure 2-9: Bubble charts of hub centrality (left) and authority centrality (right) on a directed version of Farrelly's social network

There are higher hub scores among the medical school individuals than the veterans group individuals due to the high level of interconnectivity of the former, with Farrelly's vertex V7 serving as a prominent hub in this network. Many of the medical school individuals point to V5 and V3, giving these vertices large authority scores, while V7 has a relatively modest authority score. This suggests that V3 and V5 are the primary sources of information within this community, and they pass this information along to V7, who shares it with the rest of the medical school individuals; in contrast, there is not much of a centralized structure or efficient flow of information among the veterans group's individuals.

Now that we've explored all these centrality measures visually, let's use a few of them to rank and compare the vertices in Farrelly's social network (going back to the original undirected version). The code in Listing 2-7 can be added to that of Listing 2-4 to produce a dataset of some centrality measures.

```
#create dataset of a few centrality measures on Farrelly's social network
data_social<-cbind(page_rank(g_social)$vector,degree(g_social),
hub_score(g_social)$vector,betweenness(g_social))
colnames(data_social)<-c("PageRank","Degree","Hub Score","Betweenness")
```

Listing 2-7: A script that creates a table of a few vertex centrality scores on Farrelly's social network

Table 2-1 displays these vertex scores, with the highest two entries in each column in bold.

Table 2-1: Centrality Measures Scored on the Vertices of Farrelly's Social Network

Vertex	PageRank	Degree	Hub score	Betweenness
V1	0.032	2	0.414	0
V2	0.020	1	0.168	0
V3	**0.075**	**5**	0.682	19.500
V4	0.032	2	0.370	0
V5	0.057	4	**0.685**	2.000
V6	0.046	3	0.505	1.000
V7	**0.111**	**8**	**1.000**	**100.500**
V8	0.044	3	0.550	0
V9	0.044	3	0.550	0
V10	0.048	3	0.306	**90.917**
V11	0.050	3	0.118	26.667
V12	0.022	1	0.021	0
V13	0.069	4	0.087	18.917
V14	0.067	4	0.127	61.083
V15	0.068	4	0.088	36.000
V16	0.040	2	0.024	23.083
V17	0.040	2	0.034	21.917
V18	0.065	3	0.009	20.250
V19	0.043	2	0.011	10.167
V20	0.026	1	0.002	0

By a considerable margin, betweenness identifies V7 and V10 as the most important vertices connecting the network. However, the other centrality scores for V10 are much more modest, suggesting it mainly functions as a bridge between communities rather than a true center of the network. This makes sense as V10 appears fairly undistinguished within the veterans group. It just happens to connect to Farrelly's V7 and, hence, through her to the medical school community. Farrelly's V7, on the other hand, has many neighbors (as indicated by degree) and is the top-ranked vertex for both hub score and PageRank—this is compatible with our conceptual understanding of this network as conveying two separate communities in which Farrelly is involved.

V3 is the second-highest ranked vertex in terms of degree and PageRank; this person appears to be at the social center of the medical school community. V5 has the second-highest hub score, indicating that it is close to the important members of the medical school community. Hub score (which coincides with authority here since this network is undirected) also distinguishes V12 from V20. Both of these are degree-one vertices in the periphery of the veterans group's community and with low PageRank and zero betweenness. However, V12 has a higher hub score because it is near a highly interconnected little subcommunity.

Additional Quantities of a Network

Network geometry provides more than just centrality measures for vertices. In this section, we'll introduce a few other quantities of interest associated to the vertices in a network. Examples include transitivity scores that capture how likely it is that your friends know each other, efficiency scores that capture how much traffic is diverted when vertices are removed, and curvature scores that capture distortions in the fabric of the network.

The Diversity of a Vertex

While hub and authority centrality concern the potential flow of information through a directed network, another measure associated to the vertices in a weighted network concerns information in the sense of mathematical information theory. The *diversity* of a vertex is a scaled version of the Shannon entropy (a fractional measure of information content of a variable or set of variables) of the weight distribution of the edges attached to that vertex. As the name suggests, this captures the diversity of edge weights attached to each vertex. Entropy is maximal for a uniform distribution, so the diversity score is maximal (and scaled to 1) when all the edges have equal weights, and the diversity scores decrease from this value as the weights become more varied. One way to think about this is that a random walk on a weighted network (with probabilities proportional to the edge weights) will be least predictable at vertices with diversity scores closest to 1 and more predictable for vertices with smaller diversity scores.

Triadic Closure

A useful notion in network theory is *triadic closure*, which is the tendency for triangles to form among triples of vertices that already have a pair of edges. In more down-to-earth terms, this is akin to asking whether two of your friends are likely to be friends with each other. Consider the network of friends depicted in Figure 2-10.

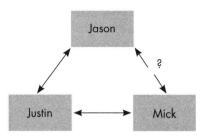

Figure 2-10: A network of friendships with one friendship link unknown, illustrating the concept of triadic closure

Here, Justin and Jason are friends, as are Justin and Mick. The question is whether Jason and Mick are friends. This situation is called a *triangle* centered at Justin. If Jason and Mick are not friends, then it is called an *open* triangle, whereas if they are friends, then it is called a *closed* triangle. A network with mostly closed triangles suggests a high degree of cohesion. In social networks, this means individuals with a mutual friend are likely to know each other. On the other hand, a network with many open triangles may indicate a less cohesive situation but could also indicate missing data and incomplete information: people may not friend all of their real-world friends on social media, intelligence data may not contain all communication forms among a cell of terrorists, and so on. In any real-world setting, relationships may exist that simply haven't been recorded yet.

One way to quantify triadic closure is through *transitivity*, which assigns to each vertex the fraction of triangles centered at it that are closed. In a social network, your transitivity is the probability that two of your friends are friends with each other. Let's add the following to our collection of plots from Listing 2-6:

```
#compute and plot transitivity for each vertex in Farrelly's social network
plot(g_social,vertex.size=20*transitivity(g_social,type="local",isolates="zero"),
vertex.label.cex=0.8,vertex.color=2)
```

Setting the parameter called type to local tells igraph to compute the transitivity around each vertex, and setting the parameter called isolates to zero forces the score for vertices of degree 1 to be 0 (which otherwise would be scored NaN, as a degree-1 vertex has no triangles centered at it). Figure 2-11 shows the resulting plot.

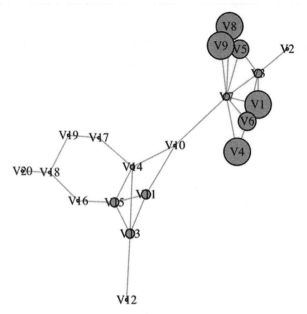

Figure 2-11: A bubble chart of transitivity on Farrelly's social network

This shows from another perspective that the medical school community is more cohesive and more tightly interconnected than the veterans' community. It's likely that medical students involved in the same sort of extracurricular activities in the same classes know each other and communicate. Note, however, that Farrelly's V7 has a relatively low transitivity score among the medical school individuals, meaning several of Farrelly's medical school friends are not friends with each other despite the overall closeness of that group. (The transitivity of V7 is also dragged down by the fact that none of Farrelly's medical school friends are friends with the veterans group's friend V10.)

The Efficiency and Eccentricity of a Vertex

Recall that two of the centrality measures we covered, closeness and betweenness, are defined in terms of the lengths of the shortest paths (or *geodesics*) in the network. Another interesting measure based on shortest paths, called *efficiency*, is defined for each vertex as follows: remove the vertex in question and then compute the resulting network distances between all pairs of neighbors of this vertex. Then, average the inverses of these distances. Removing the vertex is like creating a roadblock, and the neighbor-to-neighbor distances measure the length of the detours that must be taken because of the roadblock. Inverting these distances means bigger detours count toward smaller efficiencies, and vice versa. The efficiency measures how easily traffic can be diverted around each vertex in the network.

Efficiency is a useful way to probe the local geometry of the network. The word *local* here means we are exploring the network geometry nearby around each vertex; this is in contrast to closeness and betweenness that are more global in nature since they involve paths across the entire network. Let's add the computation of efficiency to Listing 2-6. We need an additional library to do this:

```
#load the brainGraph library that adds on to igraph
library(brainGraph)

#compute efficiency for each vertex in Farrelly's social network
efficiency(g_social,type="local")
```

We don't show the bubble chart plot for this because it looks extremely similar to the transitivity plot in Figure 2-11. In fact, the correlation between the transitivity scores and the efficiency scores for this network is 0.96. In other networks, especially larger and more complex ones, this need not be the case. It is important to note that transitivity is an extremely local measure, as it considers only neighboring vertices and the edges among them. In contrast, efficiency is local in nature but can access larger pieces of the network because the detours caused by the roadblocks may extend beyond the set of immediate neighbors.

Another path-based vertex measure that—like closeness and betweenness—sees the global geometry of the network is *eccentricity*, which assigns to each vertex the largest network distance from that vertex to any other vertex in the network. This measures how peripheral each vertex is, so lower eccentricity scores generally indicate more central vertices. We'll come back to this measure shortly.

Forman–Ricci Curvature

Ricci curvature is a concept in geometry that measures the distortion of straight lines on curved surfaces and the rate at which this distortion grows or shrinks. You can think of it as a force of sorts that warps straight lines into curved paths (like a small hand weight on a wet paper towel). See Figure 2-12 for a conceptual illustration of increasing Ricci curvature from left to right on the diagram.

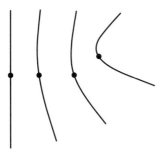

Figure 2-12: A straight line with midpoint deformed by sliding it (from left to right) through spaces with increasing Ricci curvature

One of the adaptations of this concept from continuous geometry to the realm of networks is called *Forman–Ricci curvature*. The first step in calculating Forman–Ricci curvature is to assign a number to each edge in the network that measures how spread out the network is in the immediate vicinity of the edge. The number that is used is 2 minus the sum of the degrees of the two vertices attached to the edge; this is the negated number of edges adjacent to the edge in question. For example, in Farrelly's social network (Figure 2-5), the edge between V7 and V10 has a Forman–Ricci curvature of –8 (coming from the six non-V10 neighbors of V7 and the two non-V7 neighbors of V10).

Next, we use these edge scores to assign scores to the vertices as follows: the Forman–Ricci curvature of a vertex is the sum of the Forman–Ricci curvatures of the edges attached to this vertex. Forman–Ricci curvature is almost always negative, so when visualizing it, we usually use its negation. In Listing 2-8 we compute the Forman–Ricci curvature for Farrelly's social network and then plot it using its negation for the vertex size and edge thickness.

```
#compute the degrees of all vertices
d<-degree(g_social)

#count edges and initiate vector of edge curvature values
l<-length(E(g_social))
frc<-rep(NA,l)

#loop to calculate and store Forman-Ricci edge curvature
for (I in 1:l){
  w<-as.vector(ends(g_social,E(g_social)[i],names=F))
  frc[i]<-2-d[w[1]]-d[w[2]]
}

#count vertices and initiate vector of vertex curvature values
n<-length(d)
frcv<-rep(NA,n)

#loop to calculate and store Forman-Ricci vertex curvature
for (i in 1:n){
  I<-as.vector(incident(g_social,i))
  frcv[i]<-sum(frc[I])
}

#plot the network with vertex and edge size given by the negated curvature
plot(g_social,edge.width=-frc,vertex.size=-20*frcv/max(-frcv),vertex.label.cex=0.8,
vertex.color=2)
```

Listing 2-8: *A script that computes and plots the Forman–Ricci curvature for Farrelly's social network*

Figure 2-13 shows the resulting plot.

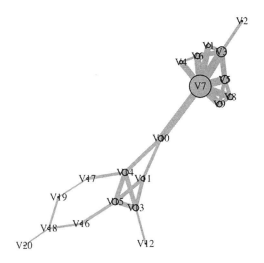

Figure 2-13: The negated Forman–Ricci curvature for Farrelly's social network

As you can see, the areas of the network with more edges have a higher negated Forman–Ricci curvature. This occurs both in hublike regions where the network fans out (like in the medical school community here surrounding Farrelly's V7) and in tightly interconnected regions (like the close-knit four-vertex subcommunity within the veterans group's members).

In traditional geometry, curvature is used to study how substances flow across objects. Similarly, Forman–Ricci curvature can be used to study network flow patterns. We'll turn to the concept of network flow in Chapter 3.

Global Network Metrics

All the metrics discussed so far are *vertex metrics*, meaning they assign a score to each vertex in a network, quantifying some aspect of the vertex's role or position within the network. For example, you could extract your network of Facebook friends and use these vertex metrics to measure how central each person is in it. However, sometimes instead of comparing vertices within a single network, we need to compare different networks. For example, you might want to compare your Facebook friend network to the Facebook friend network of someone else—or you could compare your friend network on Facebook to your friend network on a different social media platform. In this situation, we need metrics that assign a single number to an entire network; these are sometimes called *global network metrics*, or just *global metrics* if the network context is clear. Let's walk through a few important ones.

The Interconnectivity of a Network

The simplest global metrics are the number of vertices in a network and the number of edges. Closely related to these (but often more useful) is the *density*, which is the number of edges in a network divided by the maximum number of edges possible on that vertex set. An undirected network with n vertices has at most

$$\binom{n}{2} = \frac{n(n-1)}{2}$$

edges (not allowing loops or multiple edges), while the maximum number for a directed network is twice this amount. Density gives a coarse measure of how interconnected a network is; it is computed in igraph with the `edge_density()` function. Figure 2-14 shows some networks with a range of densities.

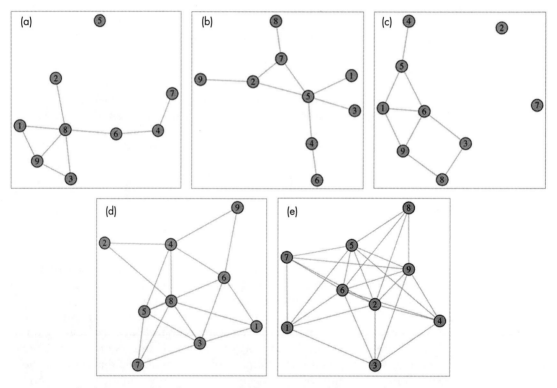

Figure 2-14: The three networks in the top row all have a density of 25 percent, while the networks in the bottom left and bottom right have densities of 50 percent and 75 percent, respectively.

The average degree in a network can also be a useful global metric, but it contains precisely the same information as the density and is perfectly correlated with it. Indeed, the average degree is the sum of all vertex degrees divided by the number of vertices, and the sum of all vertex degrees is always twice the number of edges. Thus, average degree and edge density differ by a constant that depends only on the number of vertices in the network.

Consequently, you can use one or the other measure, but nothing is gained by using both simultaneously.

The notion of transitivity discussed in the preceding section (quantifying triadic closure) has a global counterpart. When setting the type parameter to `global` in igraph's `transitivity()` function, a single number is computed that gives the fraction of triangles in the network that are closed. This gives another way, different from edge density, to measure the overall interconnectedness of a network.

Spreading Processes on a Network

Networks that are more spread out tend to behave differently than ones that are more compact. For example, consider three friends—Amara, Imani, and Taraji—sharing news with each other at a lunch table. We'd expect the girls to exchange information more easily together at a table than they would if they were in separate classes texting each other or talking as they passed each other in the hallway. Let's take a look at a few different ways to quantify the spread of a network.

Setting the type parameter to `global` in igraph's `efficiency()` function gives a global version of that metric that computes the average of the inverses of the network distances between all pairs of vertices in the network. No roadblocks are involved in this global variant of efficiency. Networks that are highly interconnected tend to have high global efficiency, as high interconnectivity means there are lots of direct routes between vertices, but, overall, it is best to think of global efficiency as a measure of the compactness of a network. The more spread out a network, the lower its global efficiency will be.

We can also compute the *diameter* of a network, which is the maximum eccentricity of the vertices in the network (and, hence, the "longest shortest path" between two vertices). The *radius* of a network is the minimum eccentricity, which measures how far the center of the network is from the most distant part of the network. In igraph, you can use the `eccentricity()` function first to score all the vertices and then use standard R commands to extract the maximum and minimum values from them. Both diameter and radius quantify some aspect of the spread of the data. One structure that can produce a large difference between the diameter and radius is a structure where there are several loose hubs connected to each other through a single bridging individual.

Spectral Measures of a Network

We have seen that eigenvectors of the adjacency matrix and other matrices closely related to it play an important role in multiple vertex centrality metrics; the spectral theory (eigenvalues and eigenvectors) of the adjacency matrix also provides us with some useful global metrics. The *spectral radius* of a network (which came up earlier in our discussion of Katz centrality) is the largest eigenvalue of the adjacency matrix. A variety of properties of the spectral radius have been established mathematically. In essence, the spectral radius measures propagation across the network. It is inversely related to the robustness of the network when considering the spread of a harmful entity across the network, such as fake news or a virus. For instance, in an epidemiological model that we'll discuss in the next chapter, a smaller spectral radius means disease spreading across the network will die out more quickly.

To compute the spectral radius of a network in R, you can use the spectrum() function in igraph that takes advantage of the sparse structure that adjacency matrices usually exhibit, or you can use any of the standard eigenvector or eigenvalue implementations in R, such as the eigen() function, on the network's adjacency matrix.

Since igraph's spectrum() function optionally returns all eigenvalues and eigenvectors of the adjacency matrix, it is easy to use it to compute other spectral measures—such as the *spectral gap*, which is the difference between the largest eigenvalue of the adjacency matrix (the spectral radius) and the second largest eigenvalue. The spectral gap controls the convergence time of certain algorithms and random processes on the network, among other things. Other useful spectral measures are based on a variant of the adjacency matrix called the *graph Laplacian*, which is obtained by negating the adjacency matrix and adding the degree of each vertex to the corresponding diagonal matrix entry. The multiplicity of the 0 eigenvalue of the graph Laplacian is the number of connected components of the network. The smallest nonzero eigenvalue of the graph Laplacian is called the *algebraic connectivity*; this is a connectivity measure of the network that conveys how difficult it is to fragment the network into smaller pieces. Farrelly's social network has a relatively low algebraic connectivity because removing a single vertex splits the network into the veterans group and the medical school group.

Listing 2-9 shows how to compute the graph Laplacian and extract from it the number of connected components and the algebraic connectivity.

```
#compute the graph Laplacian of a network g
lap<-laplacian_matrix(g)

#compute the eigenvalues and round to avoid numerical issues
evals<-round(eigen(lap)$values,digits=5)

#compute number of connected components
sum(evals == 0)

#compute the algebraic connectivity
unique(evals)[length(unique(evals))-1]
```

Listing 2-9: A script that computes the graph Laplacian of a network g and then extracts from it the number of connected components and the algebraic connectivity

For example, running this code on the two networks shown in Figure 2-15 gives seven connected components each and an algebraic connectivity of 0.062 for the network on the left and 0.09 for the network on the right.

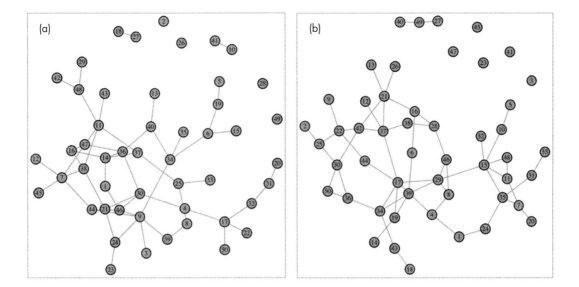

Figure 2-15: Two networks, plotted to illustrate the spectral measures associated with the graph Laplacian matrix. They have the same number of connected components, but the network on the right has a higher algebraic connectivity.

If you want to compare scores across networks of different sizes, it is better to use the normalized Laplacian; to do this, simply set normalized=T when computing the graph Laplacian in the first line of Listing 2-9.

Network Models for Real-World Behavior

There are a few different types of networks that serve as important models for real-world behavior; these provide helpful baselines against which to compare real-world networks and data. Let's start with the simplest kind to construct.

Erdös–Renyi Graphs

Erdös–Renyi graphs are networks in which edges are generated randomly according to a uniform distribution on the set of vertex pairs. These are created in igraph with the sample_gnp() function. Both the networks in Figure 2-15 were produced by using sample_gnp(50,0.05), which creates a network with 50 vertices and an edge probability of 5 percent. The density of the graphs created this way will be close to the specified edge probability but not necessarily equal to it—just as repeatedly flipping a fair coin won't always give exactly half heads and half tails. The networks in Figure 2-14 were also created with sample_gnp().

Erdös–Renyi graphs provide a useful null hypothesis. If you believe your network is highly structured, then its edges should be very far from uniformly distributed, so the network will not look or behave like an

Erdös–Renyi network. In the next two chapters, we'll see how to practically implement this idea. Erdös–Renyi graphs do not appear often in nature, especially in socially driven settings. Real networks are almost always more structured than a network with purely random edges like this.

Scale-Free Graphs

A *scale-free graph* is a network whose degree distribution asymptotically follows a power law, meaning there is a constant, c, such that the fraction of vertices of degree $2d$ is $1/2^c$ times the fraction of vertices of degree d (at least approximately, with the approximation getting more accurate as d increases). The constant c (called the *power*) usually lies between 2 and 3. This power law property leads to the existence of many vertices whose degree is much higher than the average vertex degree in the network. Consequently, scale-free networks usually have a spoke-and-wheel shape of loosely connected hubs, like airport terminals connecting at a central security gate. The hubs in the network tend to rein in the distances between vertices, giving these networks certain "small-world" properties.

These networks have a fascinating history and well-developed theory that we don't have space to get into. An interesting debate in the field has been whether many naturally occurring networks are scale-free—including internet pages, social networks, biological networks, and even airline travel networks—and if so, why this might be.

There are a variety of approaches for generating scale-free networks; one of the more popular options is conveniently implemented in igraph with the `sample_pa()` function. It relies on what is known as the Barabási–Albert model.

Watts–Strogatz Graphs

Watts–Strogatz graphs are networks generated by a random graph model introduced in 1998 that tends to produce even more small-world properties than scale-free networks, such as tightly interconnected communities and small network distances between many vertices. These networks frequently include paths that are reinforced by redundancy and alternate routes. They are common in biology and social processes. For example, brain connectivity networks at the neuron and functional area levels, voter networks, influencer networks on social media platforms, and food webs often form Watts–Strogatz networks.

Since we have already seen a few examples of Erdös–Renyi graphs (see Figures 2-14 and 2-15), let's create some networks of these other types. In Figure 2-16, we use `sample_pa(100,power=2.5,directed=F)` to generate a couple of scale-free networks, and we use `sample_smallworld(1,100,2,0.05)` to generate a couple of Watts–Strogatz networks. We leave it to a motivated reader to look in the igraph documentation to learn about the parameter choices in this latter function.

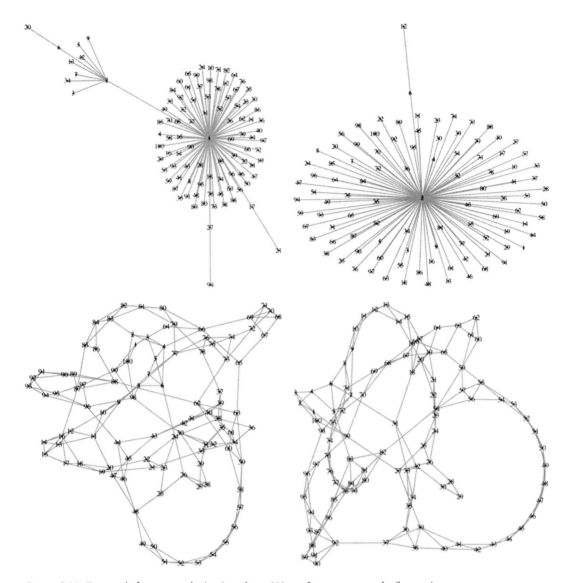

Figure 2-16: Two scale-free networks (top) and two Watts–Strogatz networks (bottom)

The scale-free networks in the top row each have one prominent hub, and the network on the left additionally has a less prominent secondary hub. Imagining these to be power grid connections, a storm that takes out a main hub would impact many more customers in the scale-free network than in the Watts–Strogatz network. This is why many planned real-world networks conform to a redundancy-heavy structure. The Watts–Strogatz networks do not exhibit strong hubs like scale-free networks, but they have a lot of structure not seen in the purely random Erdös–Renyi networks. For instance, the lengthy reinforced path structures mentioned earlier are easy to recognize in these bottom plots.

Summary

This chapter opened with a brief section explaining the need for networks in data science and the need for geometry in network science. Next, we introduced networks and the objects involved in them—vertices, edges, paths, and more. We then defined and explored a collection of metrics that quantify various properties of networks and their vertices. The chapter concluded with a few random graph models that are useful for generating synthetic network data to which real network data can be compared.

3

NETWORK ANALYSIS

In Chapter 2, we considered a few network geometry metrics; in this chapter, we'll use them. First, we'll explain how vertex metrics allow you to do supervised learning within a network, that is, predicting values associated to vertices and predicting new edges in the network; we'll also look at how vertex metrics enable you to cluster vertices in a network. We'll then discuss a few clustering algorithms that operate directly within the geometry of the network. Next, we'll explain how to use global network metrics to do machine learning and statistical analyses on datasets that consist of collections of networks. We'll then explore a network variant of the susceptible, infected, and recovered (SIR) model from epidemiology. With this model, we can see how entities (from diseases to misinformation) spread through networks and how network geometry influences this spread. Finally, we'll examine how we can use vertex metrics to devise targeted strategies for disrupting this spread.

Using Network Data for Supervised Learning

Network data is often accompanied by a traditional structured dataset where the rows are identified with the vertices in the network. For instance, you might have a social network dataset that consists of a list of individuals (the vertices), the friendships between them (edges in the network), and one or more numerical or categorical columns providing additional non-network information about each individual—such as age, gender, or salary. We might want to consider this as a supervised learning problem, where we train a machine learning algorithm to predict one of the data columns. The way to do this is to use vertex metrics as independent variables (along with any of the other data columns), which lets the algorithm incorporate the network role of each vertex when making predictions. Let's try this.

Making Predictions with Social Media Network Metrics

Let's return to Farrelly's social network, which we analyzed in the previous chapter to illustrate the different measures of vertex centrality. How might bridges between different parts of the network or hubs tightly connecting a few medical school friends influence how often Farrelly mentioned those individuals in her diary during the first term of medical school? Let's take a look at how network metrics relate to diary mentions in this social network.

We'll choose a few vertex centrality metrics and attach Farrelly's diary data as a dependent variable in the set. Let's load the data and examine the distribution of our dependent variable with Listing 3-1.

```
#import dataset
g<-read.csv("SocialNetworkModel.csv")

#create a histogram of the data
hist(g$Diary.Entries,main="Diary Entry Histogram",xlab="Diary Entries")
```

Listing 3-1: A script that imports the relevant .csv file for further analysis

Figure 3-1 shows a histogram of the dependent variable's distribution (Poisson).

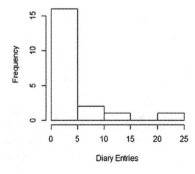

Figure 3-1: A histogram of diary entry data for Farrelly's social network

Generally, count variables, such as those conforming to the Poisson distribution, include a lot of zero and near-zero values, along with some large values. We can see from Figure 3-1 that this outcome is Poisson distributed. Poisson-distributed variables can pose issues to machine learning algorithms, as they involve a lot of zero values and some outliers, such as the diary entries involving V10. This suggests that a generalized linear model (Poisson regression) is probably more appropriate of a supervised learning model than other machine learning algorithms. It looks like most individuals in the network receive few (if any) mentions over the term. However, a few outliers exist, including Farrelly and her closest friends within the network (V3, V10, and V14). Let's dive deeper to see how centrality measures predict diary mentions. In Listing 3-2, we sample Farrelly's social network metrics of interest and use them as independent variables in our Poisson regression model.

```
#create a training sample from Farrelly's social network metrics
n<-dim(g)[1]
set.seed(10)
train.index=sample(1:n,15)
train<-g[train.index,]

#build a Poisson regression model
gl<-glm(Diary.Entries~.,data=train,family="poisson")

#examine performance with a model summary and Chi-squared test
summary(gl)

1-pchisq(summary(gl)$deviance,summary(gl)$df[2])
```

Listing 3-2: A script that computes a Poisson regression and analyzes its results

Your results may vary depending on your R version's seeding, but in the samples we modeled with this dataset and other seeds (9 of 10 random splits), the summary functions show that betweenness centrality seems to have large coefficient values in the model and be the most consistent predictor of diary entry mentions across subsets of Farrelly's social network data modeled with our regression function. The Chi-squared test values in our samples ranged from $p < 0.01$ to $p = 0.25$. When we examine the plots associated with the linear regression, we can see that most of our sample fits the regression equation well. Figure 3-2 shows two of the plots generated by Listing 3-2 (including V3, V7, and V13 as outliers).

The small sample size likely contributes to the variation between samples, but overall, we have a good predictive model. Indeed, this reflects Farrelly's own intuition that the bridges of her network tended to coordinate memorable events and activities that brought together various pieces of her network that term. We'll return to the instability of regression models on small sample sizes with outliers in Chapter 6 and Chapter 8, along with more stable models you can use for these situations to get consistent model results.

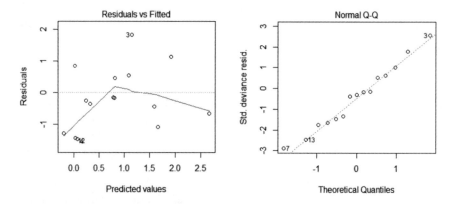

Figure 3-2: Residual and quantile plots of the Poisson regression run in Listing 3-2

Bigger networks with dependent variables more closely tied to one's social network (such as strength of political views, workout habits, and so on) tend to work better in this sort of analysis. Within biological networks, centrality measures might be used to predict disease severity, likelihood of response to a drug undergoing clinical trials, or disease risk at six months after a social-network-based behavioral health intervention. Analyzing the geometry of the network often produces useful independent variables for predicting some quality associated with the vertices of the network.

Predicting Network Links in Social Media

Another important form of supervised learning in a network is *link prediction*, in which potential new edges are inferred from a network's structure or metadata. One way to predict links is to use prior growth patterns of a network to predict which edges are most likely to appear next. This has many real-world applications, some of the most notable being in social media. Whenever Facebook or another platform suggests a person for you to friend, an algorithm has run a link prediction on its network of users behind the scenes and given the missing edge between you and this potential friend a high score. There are many sophisticated methods for performing link prediction, but a common general strategy is to translate the problem to a traditional Euclidean supervised learning task. Let's explore this conceptually using Farrelly's social network as an example.

Imagine this network evolving over time, with a binary indicator denoting edges that formed since the last time period. Wouldn't it be neat if we could predict edge formation over a time period based on what the network looked like geometrically in the past time period? Or if we could use vertex labels (such as class schedule or volunteer days) to predict edge formation in the next time period?

For the independent variables, we can use any collection of features associated with the two vertices. These can be intrinsic network-based features or extrinsic features such as user demographics in a social network. The network-based predictors come in two flavors. We can use vertex metrics by choosing a function to aggregate the two vertex scores in each pair to a single number (common choices for this include sum, max, mean, and absolute value of difference). For instance, we could compute the PageRank score for the two vertices in each vertex pair and then take the average of these two scores to assign the vertex pair.

The other flavor of network-based features uses some measure of the network relationship between the two vertices; the most natural choice here is simply the network distance between the two vertices, though you can try other options such as the number of shortest paths between the two vertices or the average time a random walk takes to get from one to the other. All these network-based predictors should be computed for the current version of the network rather than the earlier snapshot, and these network-based predictors can be combined with any collection of non-network features. (In practice, most non-network features are attached to individual vertices rather than pairs, so once again you'll have to aggregate them to get a single score for each vertex pair.)

Once the independent predictors are computed for a time period of interest and an indicator variable exists for that time period, a supervised classifier can be used to predict edge formation over the time periods of interest. The higher the likelihood score for edge formation, the more likely that relationship will exist by the next time period. In Farrelly's social network, betweenness centrality would likely be the main network-based predictor, as well as social activity data or diary entries from the first term of medical school. Everyone in her original social network was connected to her (and most to each other) by the end of that term.

Using Network Data for Unsupervised Learning

Just as we can use the vertex metrics from Chapter 2 as predictors in a supervised learning task, we can also use them as features in unsupervised learning tasks. In the case of clustering, this will partition the vertices into sets of those with similar functions in the network (hubs, bridges, and so on). Clustering vertices is known in the network sciences as *community mining*, so when using vertex metrics for this purpose, we obtain communities defined by the structural role they play in the network.

Applying Clustering to the Social Media Dataset

In Listing 3-3, we apply k-means clustering to Farrelly's social network dataset that was our main running example in Chapter 2.

```
#rescale the matrix of vertex metrics and apply k-means with k=3
clust<-kmeans(scale(vertdata),3)

#plot the network with clusters represented by vertex color and label
plot(g,vertex.size=6,vertex.color=clust$cluster,vertex.label=clust$cluster)
```

Listing 3-3: A script that uses k-means on the vertex metrics to cluster the vertices from the network in Figure 3-1 into k = 3 groups

Using k-means with $k = 3$ and PageRank, degree, hub centrality, betweenness, and transitivity as our features (which we first rescale), we get clusters with the means and sizes in Table 3-1, where they were computed from the original prescaled values.

Table 3-1: Cluster Means and Sizes for k-Means Clustering (with $k = 3$) Run on a Handful of Vertex Metrics for Farrelly's Social Network

Cluster	PageRank	Degree	Hub score	Betweenness	Transitivity	Cluster size
1	0.05	3.14	0.54	3.21	0.80	7
2	0.11	8.00	1.00	100.50	0.25	1
3	0.05	2.50	0.08	25.75	0.097	12

Since k-means involves a random initiation, you might get different clustering results each time you try this. For this particular clustering, we see that one vertex has been assigned its own cluster (Farrelly's vertex, V7); all of this vertex's scores other than transitivity are exceptionally high, so it is an outlier in many metrics. All the remaining vertices are split between the other two clusters, which seem mostly distinguished by the fact that one cluster has higher hub and transitivity scores, while the other cluster has higher betweenness. Let's plot this network with the vertices labeled by cluster (doing so is an easy adaptation of the code in Listing 3-3); see Figure 3-3.

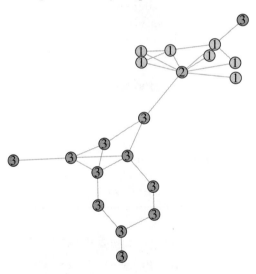

Figure 3-3: Farrelly's social network colored and labeled by cluster for the k-means clusters summarized in Table 3-1

We see that Farrelly is indeed her own cluster, and rather remarkably, cluster 1 is almost precisely the remaining medical school individuals, while cluster 2 is the veterans group individuals. Interestingly, one medical school person has been placed in cluster 3, because they are isolated (have a transitivity score of zero) and are not part of the main hub of medical school individuals.

Community Mining in a Network

So far, we've used the geometry of the network to extract features and then run traditional Euclidean machine learning clustering algorithms on them. Another approach to clustering vertices in a network (called *community mining*) is to rely directly on the geometry of the network. We'll briefly walk through several ways to do this.

Exploring Networks with Random Walks

The *walktrap algorithm* uses random walks to explore the network and find communities in which the random walk gets "trapped." If a random walk frequently stays within a certain set of vertices, then that set is a good candidate for a cluster. For instance, in Farrelly's social network, a random walk starting among the medical school individuals has a high likelihood of staying with them for many steps because the only way out is across the bridge from Farrelly's vertex. Indeed, to get out, the random walk would have to be at Farrelly's vertex, and we would then choose that bridge as the next step (which happens with only one out of eight probability from that vertex because that vertex has eight edges attached to it). Similarly, a random walk starting among the veterans group individuals has a high likelihood of staying among them. In this way, the walktrap algorithm is good at finding communities that are separated by bridges. One downside with this approach is that it is computationally intensive to explore a large network this way.

Evaluating a Cluster's Quality Outcome

In traditional Euclidean clustering, a typical way to evaluate the quality of a clustering outcome is to compare the distances within each cluster to the distances between the different clusters. There is a widely used network variant of this idea that applies to vertex clustering: the *modularity* of a vertex clustering is the probability that a randomly chosen edge is attached to two vertices in the same cluster minus the probability that this would occur if the edges in the network were randomly distributed. Intuitively, this compares the number of intracommunity edges to the number of intercommunity edges. It is sort of like a "lift" measure, because it compares how much better (in the sense of edges staying within clusters) the clustering is than a randomized benchmark. One of the important small-world properties of the Watts–Strogatz networks introduced in the preceding chapter is that they tend to be highly modular (clusters with high modularity scores).

Modularity also enables you to view vertex clustering as an optimization problem: we find the cluster division that maximizes the modularity score. It's not practical to try all possible divisions into clusters, so various algorithms have been introduced that attempt to search for high-modularity clusterings without being guaranteed to find the global optimum. Two of the popular approaches for this are *greedy algorithms*, where greedy means that at each step they go in the direction that most increases the modularity, rather than taking suboptimal steps in the short run in the hopes that they lead to greater values in the long run.

Louvain clustering is one such greedy algorithm. It starts by treating each vertex as its own cluster and then iteratively merges neighboring clusters whenever doing so increases the modularity. Once this iterative local optimization process terminates, the algorithm creates a new, smaller network by merging all the vertices that have been assigned to the same cluster. This yields a network in which each vertex is its own cluster, so the same iterative local optimization process can be run again on this smaller graph. This algorithm is quite fast in practice and tends to perform well, but it often struggles to find smaller communities within large networks. There is a faster greedy algorithm that is often used, conveniently called *fast greedy clustering*, but it tends not to reach as high modularity scores.

Understanding Spinglass Clustering

Another approach to clustering doesn't involve random walks or local optimization. Rather, it draws from statistical mechanics, a branch of physics dealing with particle interactions. *Spinglass algorithms* are based on magnetic couplings within a system of particles (positive and negative charges); they seek to optimize how the charges are aligned across the system. This can be applied to vertex clustering, called *spinglass clustering*. The basic idea is to define an energy associated to clusterings and then try to minimize this energy. This energy minimization process is usually done by *simulated annealing*, which is an algorithmic approach to optimization that also has roots in statistical mechanics. In simulated annealing, rather than always moving in the direction that decreases the energy the most (as would be done in a greedy algorithm that runs the risk of getting stuck in local optima), there is a temperature parameter that determines the probability of moving instead in a "wrong" direction. As the algorithm proceeds, the temperature is steadily lowered. This helps the algorithm explore large portions of the energy landscape early on before settling down and honing in on a particular solution. It mimics the cooling process in metallurgy in which metal purifies.

Running the Clustering Algorithms on a Social Network

Let's try running these four vertex clustering algorithms on Farrelly's social network. Listing 3-4 does this and then plots the results and computes the modularity scores (to run this, make sure you've already loaded this network data as in Listing 2-4).

```
#run walktrap, louvain, fast greedy, and spinglass clustering algorithms
cw<-cluster_walktrap(g_social)
modularity(cw) #0.505
plot(cw,g_social,vertex.size=15,vertex.label.cex=0.6,main="Walktrap")

lo<-cluster_louvain(g_social)
modularity(lo) #0.476
plot(lo,g_social,vertex.size=15,vertex.label.cex=0.6,main="Louvain")

fg<-cluster_fast_greedy(g_social)
```

```
modularity(fg) #0.467
plot(fg,g_social,vertex.size=15,vertex.label.cex=0.6,main="Fast Greedy")

sg<-cluster_spinglass(g_social)
modularity(sg) #0.505
plot(sg,g_social,vertex.size=15,vertex.label.cex=0.6,main="Spinglass")
```

Listing 3-4: A script that runs the four vertex clustering algorithms discussed earlier on Farrelly's social network, plots the results, and computes the modularity score for each

Figure 3-4 shows the resulting plots.

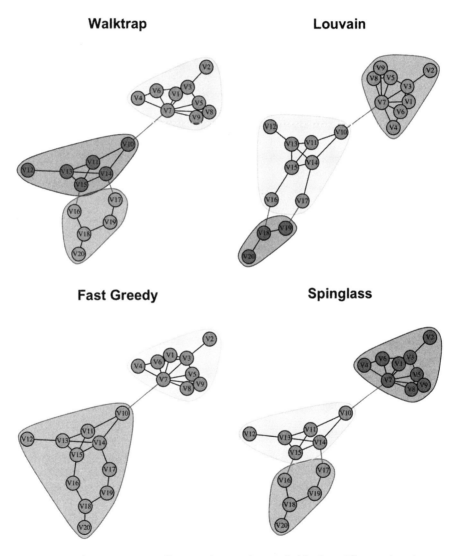

Figure 3-4: Clustering on Farrelly's social network provided by four different algorithms

Note that for all these functions the number of clusters was not specified by the user as it is for k-means. These algorithms determine the number of clusters as part of their search for optimality. In this example, walktrap and spinglass found the same three clusters, which are the medical school individuals (including Farrelly's vertex V7) and a division of the veterans group individuals into two parts. This three-way clustering yields the highest modularity score among the solutions found by these algorithms.

Louvain found the next best score, and the result is similar to the previous one except that it splits the veterans group in a slightly different way (reassigning two vertices from one cluster to the other). With a modularity score just slightly below this, the fast greedy algorithm ended up with only two clusters (the medical school community with Farrelly in it and the veterans group community). Evidently, the greediness in this algorithm prevented it from finding that a higher modularity could be achieved by splitting the veterans group community. That said, this two-cluster solution found by the fast greedy algorithm describes the original context of the data, in which Farrelly combined her two separate communities.

If you are interested, you can explore a few other vertex clustering algorithms implemented in igraph. For instance, the function cluster_edge _betweenness() uses the betweenness metric not as a feature but in a more direct way. Vertices with a high betweenness score are considered to be *bridges*, and the communities this function uncovers are the ones that are separated by these bridges. Another interesting approach is provided by the function cluster_infomap(), which uses information theory to find communities in which information flows readily; this can also be interpreted in terms of the behavior of random walks on the network.

So far, we've discussed supervised and unsupervised learning among the vertices within a single network. Let's now consider situations in which we are comparing the networks themselves.

Comparing Networks

Sometimes a network isn't your entire dataset. It's just a single instance in a dataset comprising many networks. For instance, detecting bot accounts on social media platforms usually involves supervised classification in which the friend or follower networks of real users are compared to those of fake users. In the Pennsylvania gerrymandering case mentioned in Chapter 2, districting maps were converted to networks, and the old map was shown to be a dubious outlier in the distribution of networks. Neuroscience provides another important example where one needs to compare networks. Indeed, it's common to translate functional magnetic resonance imaging (fMRI) and positron emission tomography (PET) data into a network structure in which the vertices represent different regions of the brain and in which edges are based on activity patterns (sequential activation of an area, for instance, or coactivation of multiple regions during one task). One often needs to compare two different groups of patients—either healthy patients against a group of patients with a particular neurological or psychological disorder or two different disease groups. Translated to network data science, this means we're looking at a two-class dataset of networks to see if there are statistically significant differences between the two classes (to understand structural differences). We might also want to train a supervised classifier to predict the class based on the network structure.

To generate some synthetic data, let's create 100 networks of each of the types described at the end of Chapter 2: Erdös–Renyi, scale-free, and Watts–Strogatz. In Listing 3-5, we do this and plot a histogram of the network diameter to see how it varies within and across the different types of networks.

```
#initiate vectors/lists
n<-100
er<-list()
sf<-list()
ws<-list()
er_d<-rep(NA,n)
sf_d<-rep(NA,n)
ws_d<-rep(NA,n)

#loop to create and store random graphs and compute their diameters
for (i in 1:n){
  er[[i]]<-sample_gnp(100,0.02)
  sf[[i]]<-sample_pa(100,power=2.5,directed=F)
  ws[[i]]<-sample_smallworld(1,100,1,0.1)
  er_d[i]<-diameter(er[[i]])
  sf_d[i]<-diameter(sf[[i]])
  ws_d[i]<-diameter(ws[[i]])
}

#plot combined histogram
hist(er_d,col=rgb(0,0,1,0.2),xlim=c(0,max(max(ws_d),max(ws_d),max(ws_d))),ylim=c(0,40),
xlab="Diameter",main="")
hist(sf_d, col=rgb(0,0,1,0.5), add=T)
hist(ws_d, col=rgb(0,0,1,0.8), add=T)
box()
```

Listing 3-5: A script that generates 300 networks, evenly split among three different types; computes their network diameter; and then plots the histograms for each: Erdös–Renyi, scale-free, and Watts–Strogatz

The parameters are chosen here so that all random networks have the same number of vertices (chosen arbitrarily to be 100) and approximately the same edge density (around 2 percent); this ensures that the network structure does differentiate the three groups, rather than something simpler like the number of vertices or edges. Figure 3-5 shows the resulting histogram plot.

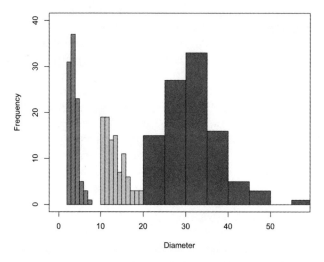

Figure 3-5: Histograms of network diameter for three different types of random networks: Erdös–Renyi (light gray), scale-free (medium gray), and Watts–Strogatz (dark gray)

We see that the three histograms are disjoint. Erdös–Renyi networks have moderate diameters. Scale-free networks have small diameter values. Watts–Strogatz networks have large diameters. If interested, you might try modifying Listing 3-5 to compute some of the other global network metrics discussed in Chapter 2 (such as efficiency, transitivity, and spectral radius) to see how they behave for the different types of random graph structures.

There are many machine learning tasks that you can do now on datasets of networks using the tools developed so far. For classification (such as labeling social media accounts as bot versus real based on their friend network) or regression (such as predicting the journal ranking of academic publications based on their citation networks), you can compute a collection of global network metrics to then use as features for a traditional supervised learning algorithm. Similarly, to cluster a collection of networks into different types (for example, grouping individuals according to their fMRI brain network structure), you can compute global network metrics and feed them into a traditional clustering algorithm. You can also do statistics, such as outlier detection and confidence interval estimation. Indeed, by representing each network with its vector of global network metric values, you "structure" your network data and open the door to all the statistical and machine learning methods that we traditionally rely upon in data science.

Analyzing Spread Through Networks

Another important topic in network analysis is the spread (or *propagation*) of various entities through a network. There are many real-world instances of this, including infectious diseases in contact networks and viral content in social media networks. Understanding the geometry of a network can help predict the way that entities spread on the network, and we can leverage this insight to change the network geometry so that we impact spread.

Tracking Disease Spread Between Towns

Let's return to the weighted network of four towns from the previous chapter's Figure 2-3. We'll take the adjacency matrix for it from Listing 2-3 and use it to create a weighted network whose edge weights are the inverses of the original distance; this turns distances into proximity scores, where shorter roads have larger edge weights than longer roads. Listing 3-6 (which relies on first running the script in Listing 2-3) does this and plots the result.

```
#invert the nonzero entries in the towns adjacency matrix from last chapter
townprox<-apply(towns,MARGIN=c(1,2),function(x) 1/x)
townprox[which(townprox == Inf)]<-0

#create weighted network from this new adjacency matrix
g_townprox<-graph_from_adjacency_matrix(townprox,mode="undirected",weighted=T)

#plot network with edges labeled by weights
plot(g_townprox,edge.label=round(E(g_townprox)$weight,3),vertex.color=2,vertex.size=15,
vertex.label.cex=0.8)
```

Listing 3-6: *A script that creates and plots the four towns network from Chapter 2 but with the edge weights given by the inverses of the road lengths*

Figure 3-6 shows the resulting plot.

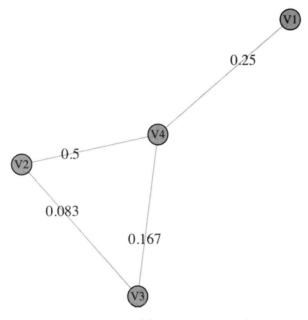

Figure 3-6: *Four towns and the proximity scores (inverse distance) of the roads between them*

Let's consider a simple epidemiological model for the spread of a transmissible disease across this network in which the probability of the disease spreading from an infected town to each of its neighboring towns is given by the edge proximity scores. If the disease starts in town V1, then it has a 25 percent chance of spreading to V4. If it does this, it then has a 16.7 percent chance of further spreading from V4 to V3. But multiplying these two probabilities (which yields about 4.2 percent) does not give the probability that the disease spreads from V1 to V3; it gives only the probability that it does so along the transmission route V1→V4→V3. Another potential transmission route is V1→V4→V2→V3, which has a 1 percent chance of occurring.

For a larger network, computing all the conditional probabilities based on potential transmission routes given by paths in the network will clearly be too cumbersome to do by hand, so we need these calculations to be automated. Moreover, this epidemiological model is too simple to be of much practical use; we discussed it here just to give a sense of how the structure of a weighted network might influence the spread of various entities (disease, information, and so on) across its vertices, as well as to motivate the more sophisticated epidemiological model that we'll be turning to next.

An *SIR model*, or *susceptible-infected-resistant* model (alternatively, a *susceptible-infected-recovered* model), is a model that projects the spread of a disease among a population by assuming each individual can be in one of three disease states: susceptible (can become infected), infected (has the disease and can transmit it), or recovered or resistant (immune to the disease). Many variations of this model exist, including models that are susceptible-infected-susceptible, models that are susceptible-infected-recovered-susceptible, models including partial immunity from vaccines, models where individuals are born or die during the epidemic, and partitioned or geographic models where populations mix at different rates. Underlying all these models are systems of partial differential equations with parameters related to population mixing (such as contact rates or times) and disease characteristics (such as the number of new infections expected for a single infectious individual). Often these differential equations are too difficult to solve explicitly, so instead we run computer simulations to quantify the range and likelihood of different possible outcomes.

To take into account the social interactions between individuals within a population, SIR models have been adapted to networks. This provides more detailed predictions of how a disease might spread, and it also helps people find ways of mitigating this spread: we can run the model to see what impact deleting a vertex or edge, or restructuring the network in other ways, will have on the spread of the disease.

Many of the network geometry concepts from Chapter 2 play a role here. Hubs are high transmission zones that might need to be shut down or reduced in size, bridges and vertices with high betweenness scores suggest targeted ways of cutting off the main transmission routes of the disease, and vertices with high centrality scores might indicate the individuals most important to vaccinate or quarantine as quickly as possible. Moreover, SIR models on networks and the computer simulation techniques used to explore them have applications far beyond epidemiology because they give a powerful empirical method for studying the complex relationship between network structure and network propagation more generally. For instance, the spread of misinformation on social media is a problem that has attracted a lot of attention recently and driven a need to better understand how network structure influences social media virality—and SIR-type models have proven to be valuable tools in this realm.

Tracking Disease Spread Between Windsurfers

Let's jump right in with an example. Listing 3-7 loads a popular network dataset, the KONECT Windsurfer Network. This is a weighted network representing 43 Southern California windsurfers and their level of interactions during the fall of 1986. Almost all the nondiagonal entries of the adjacency matrix are nonzero—meaning almost every possible edge exists in this network—so it's really the weights that matter. This makes it difficult to visualize the network, so let's create two less dense versions of the network—one with all the edges whose weights are not in the top quartile removed and one with those below the median removed. (This is a simple form of filtering weighted networks, a concept we'll return to in more depth in Chapter 4.) Listing 3-7 does this and plots the results.

```
#load dataset, compute quartiles, and convert to weighted network
wind<-as.matrix(read.csv("beachdata.csv",header=F))
q<-quantile(wind,prob=c(.25,.5,.75))
g_wind<-graph_from_adjacency_matrix(wind,mode="undirected",weighted=T)

#new networks, keeping only edges with weight in top one and two quartiles
wind_top<-wind
wind_top[which(wind < q[3])]<-0
g_wind_top<-graph_from_adjacency_matrix(wind_top,mode="undirected",weighted=T)
wind_mid<-wind
wind_mid[which(wind < q[2])]<-0
g_wind_mid<-graph_from_adjacency_matrix(wind_mid,mode="undirected",weighted=T)

#plot these two thinned-out networks with weights^2 as edge thickness
#(squaring the weights is just to increase the visual distinction)
plot(g_wind_top,vertex.size=10,vertex.label.cex=0.4,vertex.color=2,
edge.width=E(g_wind_top)$weight^2)
plot(g_wind_mid,vertex.size=10,vertex.label.cex=0.4,vertex.color=2,
edge.width=E(g_wind_mid)$weight^2)
```

Listing 3-7: A script that loads the KONECT Windsurfer Network dataset and creates two less dense versions of it, by removing edges whose weights are not in the top one or two quartiles, and then plots the result

Using the edge_density() function, we find that the original network has a density of 99.3 percent, the top-quartile network has a density of 25.8 percent, and the above-median network has a density of 51.4 percent. The plots in Figure 3-7 show these two thinned-out versions of the windsurfer network. The edge thicknesses represent the edge weights, but to increase the visual distinction among them, we set the thickness to the square of the edge weight.

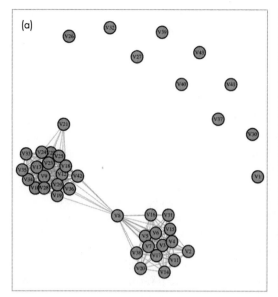
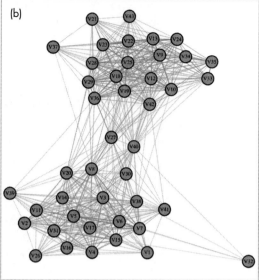

Figure 3-7: Two thinned-out versions of the windsurfer network, in which all edges whose weights are not in the top quartile (left) or top two quartiles (right) have been removed

Running an SIR simulation in R is easy. Using igraph's sir() function, you can just specify the network, the infection rate (called *beta*), and the recovery rate (called *gamma*), and then (optionally) specify the number of simulation trials to conduct (the default value is 100). The infection rate determines the probability at each time step that a susceptible vertex becomes infected by an infected neighbor (higher rates for more contagious diseases); having two infected neighbors doubles the odds of getting infected. The recovery rate determines the probability distribution for the duration of infections; higher recovery rates mean higher probability at each time step that infected vertices move on to the recovered state. Higher recovery rates indicate shorter-duration infections.

When plotting the result of sir(), you'll see the number of actively infected individuals in the network together with the median value across the trials and estimated confidence intervals as a function of time. Applying the function median() to the output of sir() provides three time series: the median number of susceptible individuals, the median number of infected individuals, and the median number of recovered individuals. Let's try this. In Listing 3-8 we simulate a disease on the full dataset with an infection rate of 3 and a recovery rate of 2.

```
#SIR simulations on the original windsurfer network
sim<-sir(g_wind,beta=3,gamma=2)

#plot the result
plot(sim,main="Number of Infected Over Time, Including Confidence Intervals")
```

```
#display the median number of infected individuals for each time bucket
median(sim)$NI
```

Listing 3-8: A script that runs 100 simulation trials of an SIR model on the KONECT Windsurfer Network dataset with an infection rate of beta=3 and a recovery rate of gamma=2 and then plots the results and displays the median number of infected individuals across time

Figure 3-8 shows the resulting plot.

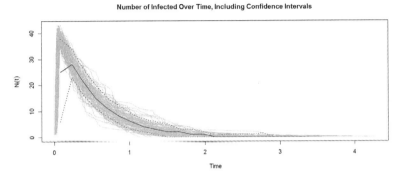

Figure 3-8: Plot of SIR simulations (100 trials) on the original KONECT Windsurfer Network dataset, showing the number of infected individuals over time (with mean and confidence intervals) for a disease with infection rate 3 and recovery rate 2

At its peak, the median is 28 actively infected individuals—which is 65 percent of the entire network. This is a highly infectious disease spreading through a densely connected network. Running the same code as in Listing 3-4 but lowering the infection rate to beta=1 and raising the recovery rate to gamma=10 yields the plot in Figure 3-9.

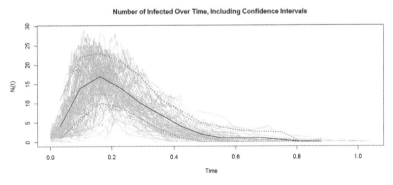

Figure 3-9: SIR simulations on the original KONECT Windsurfer Network dataset (with confidence levels), now with infection rate 1 and recovery rate 10

As expected, this new epidemic simulation shows fewer infected individuals over a smaller period of time. Now the median number of active infections peaks at 17, and the time period of this epidemic is only one-quarter what it was for the previous parameters. When using SIR models to study real-world epidemics, epidemiologists look up the infection and recovery rate parameters in the scientific literature if they are already known, and if they're not already known, they can be estimated from data on how the disease has spread so far. Usually, such estimates will involve some degree of uncertainty, so we can run SIR simulations on a range of parameters to see the range of possible outcomes.

Disrupting Communication and Disease Spread

One of the interesting proposed uses of vertex Forman–Ricci curvature is to rank vertices for removal to disrupt communication and disease spread on a network. In a communications network, disrupting communication may involve targeting a specific cell tower or isolating an individual import to the network. In 2020, we saw how isolating COVID-infected or COVID-exposed individuals by social distancing and quarantines helped stop the spread of COVID in large cities. Recall that vertex 7 in the author's network had a large Forman–Ricci curvature. Let's run an SIR model on the author's network with and without vertex 7 included to compare the results:

```
#run and plot SIR epidemic on full author's network
sim1<-sir(g_social,beta=3,gamma=2)
plot(sim1,main="Epidemic on Full Author's Network")

#remove vertex 7 from the author's network and rerun SIR epidemic

g2<-delete_vertices(g_social,v=7)
sim2<-sir(g2,beta=3,gamma=2)
plot(sim2,main="Epidemic on Author's Network with Vertex 7 Removed")
```

This script runs epidemics on the original network and the modified network, with vertex 7 removed, to compare the severity of the simulated epidemic. This should yield an initial plot similar to Figure 3-10, with the epidemic propagating through the whole network.

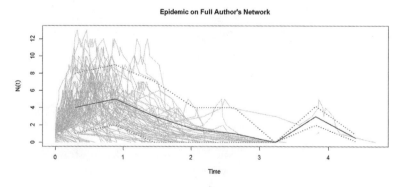

Figure 3-10: An SIR epidemic on the author's full network

Figure 3-10 shows an SIR epidemic resulting in five time periods of infection spread, with a median number of infected at 5. Some simulations suggest the possibility of up to 12 infections at once within the first 2 time periods. This is a pretty severe epidemic, forecast to impact more than 25 percent of the population at a time before the infection takes out the whole susceptible population.

Let's examine what happens when vertex 7 is removed, shown in Figure 3-11.

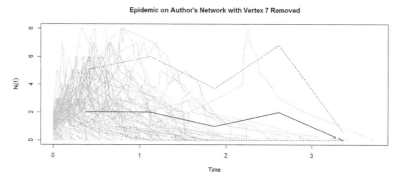

Figure 3-11: An SIR epidemic on the author's edited network (with vertex 7 removed)

Figure 3-11 shows a less severe epidemic over a shorter time frame. There are only four periods in which infection occurs, and the median number infected at the height of the epidemic is only 2, with a maximum estimate of 8. While the epidemic still impacts the population, it is confined to fewer individuals and quickly over. By the end of the second time period, most models suggest the epidemic has ended.

Many applications that let us target vertices to disrupt a network exist. Not only can we mitigate potential epidemics by removing pieces of a network, but we can also disrupt enemy communication within a terrorist cell or hostile government by taking out targets with highly negative Forman–Ricci curvature or disrupt disease processes by taking out proteins or genes within the backbone of the biological network. Changes in Forman–Ricci curvature as a network evolves also contribute to network-related analytics capabilities.

Forman–Ricci flow is a geometric flow (differential equation) related to changes in curvature over time on a network, analogous to heat dissipating across a network from a defined starting point. Tracking changes in curvature can identify areas of change within a network. Forman–Ricci flow provides a way to quantify regions of growth or shrinkage of connections within networks, such as the rapid expansion of terrorist cell membership, accumulation of mutations in a cancer gene network, or increased disease spread risk in an epidemic. For instance, the increased large party or event activity in some regions during COVID quarantines resulted in an increased spread of COVID within those parts of an area's social network. Forman–Ricci flow on image datasets also provides a way to map medical image data from a raw source file onto a standard surface, such as a plane or a sphere or even a frying pan, such that results can be pooled and compared within and across patient groups.

Summary

In this chapter, we first saw how the vertex metrics covered in Chapter 2 can serve as predictors for supervised learning within a network (including link prediction) and as features for vertex clustering (that is, community mining). This approach to machine learning translates a network back to structured datasets and applies Euclidean machine learning algorithms. We then looked into a handful of community mining algorithms that operate directly within the network. Next, we moved from analyzing data within a network to analyzing datasets where each data point is itself a network. Similar to our use of vertex metrics in the previous setting, here we used global network metrics as predictors or features to do machine learning and statistical analyses on this kind of network data. Finally, in the last section of this chapter, we explored the SIR disease spread model from epidemiology as it applies to networks. The emphasis here is on the intersection of network geometry and network spread; in particular, we discuss some targeted strategies for disrupting epidemic spread that are rooted in network geometry.

4

NETWORK FILTRATION

We've explored many ways to analyze network data by measuring geometric properties. In this chapter, we'll introduce network filtration for weighted networks, which tracks geometric properties and network metrics over threshold values imposed on the network. Then we'll examine how network data can be transformed into a higher-dimensional topological object called a *simplicial complex,* and we'll explore higher-dimensional versions of the network metrics we've previously considered. From there, we'll return to graph comparisons using a tool from topology related to filtrations.

Graph Filtration

In the previous chapters, we reviewed different network metrics, including different measures of centrality, entropy, spectral radius, diameter, and many others. There's an interesting way to understand topological properties of weighted networks: *graph filtration*, a method of creating a series of weighted networks by iteratively removing edges below a certain threshold (for instance, all edges with weights lower than 0.2, 0.4, or 0.6). By creating a series of thresholded graphs, it's possible to identify persistent network metrics, or local and global network metrics that persist across a wide range of filtration values. This gives us features that can be plotted or tracked across filtrations. This is one of the core ideas of topological data analysis (TDA).

To explore this further, let's say we're examining longitudinal educational or risk behavior outcomes of adolescents based on adolescent friendship or informal social ties within a community. Imagine we have weighted social networks with high degree metrics for each vertex, where edges are weighted by hours spent with friends over a normal week. The first group of friends might spend a couple of hours together playing soccer on the weekend. The second group might study together once or twice a week and see each other in classes. The third group might play sports often, do homework together after dinner or in the mornings before school, and stay over at each other's homes often. As we filter hours spent together, the degree metrics will drop for the first two groups of friends in a network. The last group will retain a high degree metric over the filtration, as they spend more time together. This persistence of degree will likely shed light on the strength of whatever social ties we're examining in our study.

Let's examine how we can implement graph filtrations by decomposing and exploring two small example social networks, Graph 1 and Graph 2. First, we'll load the two networks into R and explore the structures of the full networks with the script in Listing 4-1.

```
#load both networks in R
mydata1<-as.matrix(read.csv("Graph1w.csv",header=F))
mydata2<-as.matrix(read.csv("Graph2w.csv",header=F))

#load igraph and convert to graph objects
library(igraph)
g1<-graph_from_adjacency_matrix(mydata1,mode="undirected",weighted=T)
g2<-graph_from_adjacency_matrix(mydata2,mode="undirected",weighted=T)

#plot the two graphs
plot(g1,edge.label=E(g1)$weight,main="Graph 1")
plot(g2,edge.label=E(g2)$weight,main="Graph 2")
```

Listing 4-1: A script that loads two different network structures for filtration

The script in Listing 4-1 should load two different networks, Graph 1 and Graph 2, which have different connectivity patterns but the same number of vertices. It should also plot both networks with edge weights given in the plots. Let's compare the networks, shown in Figure 4-1.

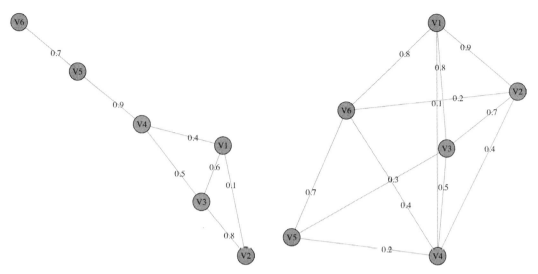

Figure 4-1: Plots of the two example networks

Figure 4-1 suggests that Graph 1 is a sparsely connected network with mostly large edge weights (perhaps a sample of students in the same class showing up for a service activity over the course of a weekend), whereas Graph 2 is a densely connected network with a mixture of different edge weights (perhaps a friendship network within a sports team). We'd expect higher hub scores and other centrality measures in Graph 2, but a filtration might change those metrics more quickly than we'd expect them to change in Graph 1.

Let's create filtrations of the networks; this will allow us to explore a few centrality metrics on these networks. We can do this by adding the following code to the script in Listing 4-1:

```
#filter Graph 1
mydata1[mydata1<0.2]<-0
g12<-graph_from_adjacency_matrix(mydata1,mode="undirected",weighted=T)
mydata1[mydata1<0.4]<-0
g14<-graph_from_adjacency_matrix(mydata1,mode="undirected",weighted=T)
mydata1[mydata1<0.6]<-0
g16<-graph_from_adjacency_matrix(mydata1,mode="undirected",weighted=T)
mydata1[mydata1<0.8]<-0
g18<-graph_from_adjacency_matrix(mydata1,mode="undirected",weighted=T)

#filter Graph 2
mydata2[mydata2<0.2]<-0
g22<-graph_from_adjacency_matrix(mydata2,mode="undirected",weighted=T)
mydata2[mydata2<0.4]<-0
g24<-graph_from_adjacency_matrix(mydata2,mode="undirected",weighted=T)
mydata2[mydata2<0.6]<-0
g26<-graph_from_adjacency_matrix(mydata2,mode="undirected",weighted=T)
mydata2[mydata2<0.8]<-0
g28<-graph_from_adjacency_matrix(mydata2,mode="undirected",weighted=T)
```

The previous code filters Graph 1 and Graph 2 by edge weight, using increasing intervals of 0.2. This yields a series of five networks in each graph filtration, which can be further examined by applying network metrics to each sequence of filtered graphs.

Let's examine the degree centrality of each vertex across the filtration of Graph 1 by adding the following to our script:

```
#calculate degree centrality for Graph 1's filtration sequence
d1<-degree(g1)
d12<-degree(g12)
d14<-degree(g14)
d16<-degree(g16)
d18<-degree(g18)

#create a dataset tracking degree centrality across the filtration
g1deg<-cbind(d1,d12,d14,d16,d18)
```

This code calculates degree centrality across filtrations of Graph 1, which should yield a dataset containing the information in Table 4-1.

Table 4-1: Degree Centrality Across Graph 1 Filtrations

Column1	d1	d12	d14	d16	d18
V1	3	2	2	1	0
V2	2	1	1	1	1
V3	3	3	3	2	1
V4	3	3	3	1	1
V5	2	2	2	2	1
V6	1	1	1	1	0

Table 4-1 shows that vertices 1, 3, and 4 have high degree centralities; however, vertices 3 and 4 retain these high degree centrality values across much more of the filtration than vertex 1, suggesting they are more important to the network, despite having the same centrality metric on the unfiltered network (column 1).

Now, let's add some code to calculate degree centrality across Graph 2's filtration:

```
#calculate degree centrality for Graph 2's filtration sequence
d2<-degree(g2)
d22<-degree(g22)
d24<-degree(g24)
d26<-degree(g26)
d28<-degree(g28)

#create a dataset tracking degree centrality across the filtration
g2deg<-cbind(d2,d22,d24,d26,d28)
```

This code calculates degree centrality across the filtration of Graph 2, yielding a table similar to that obtained by Graph 1's filtration and centrality calculation. Table 4-2 summarizes the findings from the Graph 2 filtration and centrality calculation.

Table 4-2: Degree Centrality Across Graph 2 Filtrations

Column1	d2	d22	d24	d26	d28
V1	4	3	3	3	3
V2	4	4	3	2	1
V3	4	4	3	2	1
V4	5	4	3	0	0
V5	3	3	1	1	0
V6	4	4	3	2	1

As Table 4-2 shows, there are relatively high degree centrality measures in the unfiltered Graph 2; however, the pattern changes by vertex after the filtration begins. Some vertices, like vertex 1, retain a high degree centrality throughout the filtration. Others, such as vertex 4, retain a high degree centrality and then drop to 0. Others still, like vertex 6, show a slow degradation of degree centrality over the full filtration. This may be informative in a study of social ties within a subgroup of interest. A high degree of informal social ties, represented by a high centrality degree, has been linked to positive educational attainment, career achievement, and resilience to life adversity in young adults.

Degree centrality is only one example of metrics that we can calculate across a filtration; we can also calculate other local metrics such as betweenness centrality or triadic closure. In addition, we can calculate global metrics, such as the spectral radius or the Euler characteristic, across a filtration. Let's add the following to Listing 4-1 to calculate the diameter of each filtration of Graph 1:

```
#calculate graph diameter of Graph 1's filtration
di1<-diameter(g1)
di12<-diameter(g12)
di14<-diameter(g14)
di16<-diameter(g16)
di18<-diameter(g18)
```

The sequence of diameters calculated across the filtration of Graph 1 by this code is 2.1, 2.9, 2.9, 1.6, and 0.9. Let's calculate the diameters for Graph 2's filtration:

```
#calculate graph diameter of Graph 2's filtration
di2<-diameter(g2)
di22<-diameter(g22)
di24<-diameter(g24)
di26<-diameter(g26)
di28<-diameter(g28)
```

The sequence of diameters calculated across the filtration of Graph 2 by this code is 0.9, 1.2, 1.6, 2.4, and 1.7. This is different than Graph 1's diameter sequence, suggesting that the diameter is generally smaller until later in the filtration sequence. This metric's filtration might be useful in assessing a community's overall level and depth of informal social ties, a measure of community resources available to residents in need. Figure 4-2 shows the diameter plots across both filtrations to compare the two networks.

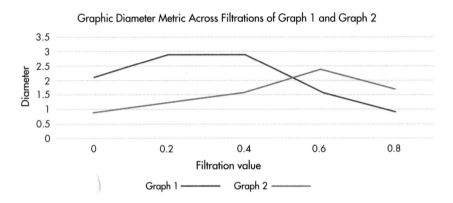

Figure 4-2: A plot of graph diameter metrics across filtrations of Graph 1 and Graph 2

As we can see in Figure 4-2, Graph 1 has a larger graph diameter than Graph 2 early in the filtration, but this relationship switches after a filtration value of 0.4. This suggests that there is greater eccentricity in Graph 1 early in the filtrations but greater eccentricity in Graph 2 later in the filtration. Remember that eccentricity is the maximum distance from one point to another in the network.

Graph filtration tracking as we've plotted in Figure 4-2 can be helpful in distinguishing similar graphs with different connectivity patterns or weights. Dynamic networks, in which weights can change over time, could be a use case of graph filtrations. In addition, they are quite useful in comparison among networks with the same vertices but potentially different weights (such as patient groups in brain imaging studies); in fact, brain imaging studies are one of the applications for which graph filtration was developed. Higher eccentricity values suggest longer pathways to relay neural signals; stronger edge weights represent stronger connections between two areas of the brain. Strong edges with low eccentricity suggest a functional module activated in a particular task given to the patient groups on which imaging was performed.

Although graph filtration is a relatively new concept, it has mainly been confined to biological network data, including networks based on brain imaging studies. However, the graph filtration method is widely applicable to weighted network data, and its tool set lends itself to further development in other fields. If you want to explore this topic in more depth, look through the references at the end of this book and play around with graph filtrations on their own data. For now, let's turn our attention to a topological view of graphs, which allows us to extend the relationships captured in graphs to other types of interactions between people or things.

From Graphs to Simplicial Complexes

Graphs can be considered topological objects that have defined global properties we can leverage in our analyses, and it's possible to turn a graph into a higher-dimensional version of a graph, called a *simplicial complex*, by considering three-way, four-way, and n-way interactions by individuals and vertices in the graph. Let's consider three colleagues who often collaborate on academic papers but have never published with all three names on a paper. We'll create a simple graph for the three colleagues, shown in Figure 4-3.

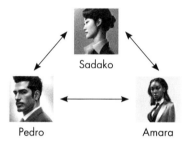

Figure 4-3: A simplicial complex showing two-way interactions among three colleagues

Now let's imagine a paper where all three colleagues participate and have their names on the paper. This is a three-way interaction, rather than three two-way interactions, and we'd end up with a filled-in triangle rather than three sets of two-way arrows, as shown in Figure 4-4.

Figure 4-4: A simplicial complex showing three-way interactions among three colleagues

Figure 4-4 uses a triangle to represent a three-way connection among colleagues, similar to how the arrows between two colleagues represented two-way connections. This can be generalized to tetrahedra for four-way interactions and more exotic shapes to represent higher n-way interactions. There's no limit as to how high of a number n can be, but computational issues will come into play at some point as we work our way up to n-way interactions in a simplicial complex. Analyses involving email chains, co-authors on papers, or conference calls are common applications that extend social network analysis and graphs into the analysis of simplicial complexes. Depending on the size of the network and the size of the n-way interactions, simplicial complex representations of individuals and mutual interactions can become very complicated across values of n. Analyzing these structures can involve a lot of computing power and tools that extend network metrics. However, because graphs are topological objects, many theorems and tools of topology can be successfully applied to them without transformations or other hassles. This, in turn, allows for other areas of math, including partial differential equations and probability theory, to be applied and developed on graphs.

Just as we could filter a weighted graph, we also can filter simplicial complexes. The filtration process for simplicial complexes varies depending on how the simplicial complex is built. In most topological data analysis algorithms, we start with a point cloud of data within a space where a distance metric can be defined. Points are included in a simplicial complex if they share either mutual n-way overlapping sets with each other (*Čech complex*) or pairwise overlapping sets (*Vietoris–Rips complex*). By sequentially increasing or decreasing the value of the distance metric, we obtain a filtration of simplicial complexes. In practice, the Vietoris–Rips complex is easier to compute and underlies many common topological data analysis packages. This leads us to a very new and emerging part of network analytics: extensions of network tools to simplicial complexes.

Many of the tools introduced in the previous chapters have simplicial complex analogs, including eccentricity, shortest path algorithms, centrality metrics (Katz centrality, eigenvector centrality, closeness centrality, and so on), triadic closure, and many more. Typically, simplicial complexes of network data are built by computing maximal cliques within the network (though it's possible to define a distance metric and apply the process defined in the prior paragraph to build simplicial complexes from network data as well). *Maximal cliques* of a network include the highest n-way mutual edges among groups of vertices. These maximal cliques correspond to an $(n-1)$-simplicial complex. The *flag complex* of the graph involves building the graph's simplicial complex by computing the graph's maximal cliques. From this complex, it's possible to define quantities at each simplicial complex level, which can be combined into a total metric across levels. This means we can glean more information about the overall structure of the network and its components at various levels of a simplicial complex.

Let's return to Farrelly's social network introduced in prior chapters and look at an extension of degree centrality, dubbed *topological dimension*. We can define topological dimension as a weighted degree centrality, weighting each vertex by the dimension of the cliques in which it resides, which involves summing across a vertex's cliques of different dimensions. For instance, a vertex in a maximal two-clique and a maximal three-clique within the network would have a topological dimension of 5. A vertex in a maximal five-clique and no other cliques would also have a topological dimension of 5. However, the former vertex might have a degree of 3, connecting to one other vertex in the two-clique and two other vertices in the three-clique; the latter would have a degree of 4, connecting to the four other vertices in the five-clique.

In Listing 4-2, we have a script that calculates the maximal cliques and the topological dimension of vertices within Farrelly's social network.

```
#load the author's network
g_social<-read.csv("SocialNetwork.csv")

#create the graph
library(igraph)
g1<-graph_from_adjacency_matrix(g_social,mode="undirected",weighted=F)

#compute the maximal cliques in the author's network data
cl<-maximal.cliques(g1)

#create array
cl<-as.array(cl)

#get clique size from maximal clique array
d<-dim(cl)
l<-rep(NA,d)
for (i in 1:d){
  l[i]<-length(as.vector(cl[[i]]))
}

#create matrix of vertices in maximal cliques
av<-matrix(rep(NA,d*20),20)
for (i in 1:20){
  for (j in 1:d){
    av[i,j]<-i%in%cl[[j]]
  }
}

#convert to binary indicators
avind<-ifelse(av==TRUE,1,0)

#multiply out to calculate each vertex's topological dimension
topmat<-t(avind)*l
topdim<-colSums(topmat)
```

Listing 4-2: A script that calculates topological dimension across vertices in Farrelly's social network

This script results in a topological dimension calculation based on the flag complex of the graph. It first calculates the flag complex from the maximal cliques; it then stores the information of each clique, such that we can cycle through each clique to see which vertices belong to each clique. Converting this information to a binary indicator matrix allows us to multiply the dimension of the clique and the indicator matrix, resulting in a vector containing the topological dimension of each vertex. Table 4-3 shows the topological dimension and degree of each vertex in the author's network dataset.

Table 4-3: Topological Dimension and Degree Summary for Vertices in Farrelly's Social Network

Vertex	Degree	Topological dimension
1	2	3
2	1	2
3	5	11
4	2	3
5	4	7
6	3	6
7	8	18
8	3	4
9	3	4
10	3	6
11	3	5
12	1	2
13	4	8
14	4	7
15	4	8
16	2	4
17	2	4
18	3	6
19	2	4
20	1	2

Table 4-3 shows a distinct difference between degree, which includes only the vertices and edges of the author's network in its calculation, and the topological dimension, which includes higher-order interactions. For instance, vertices 9 and 10 both have a degree of 3; however, their topological dimensions differ, with vertex 9 having a score of 4 and vertex 10 having a score of 6. The importance of vertex 10 to the overall network structure is larger than the importance of vertex 9 to the overall network structure. Without considering higher-order interactions within the network, we would not be able to distinguish between the two vertices with respect to this metric.

For weighted networks, it's possible to combine these simplicial-complex-based metrics with graph filtration, yielding a sequence of metrics over the filtration based on the simplicial complex of the network. You'll see this when we discuss a tool called *persistent homology* in the next section of this chapter. You could do the same with the Euler characteristic or the topological dimension or a yet-to-be-developed simplicial complex extension of network metrics.

Simplicial complex extensions of network metrics are a very new area of study within network science, and few packages or open source functions exist to calculate the simplicial analogs of network metrics. However, it is hoped that this example and some of the papers on this topic will spark the addition of simplex-based metric within network science packages. Perhaps you will take up the challenge and contribute functions to the igraph package or other open source network science tools.

The next tools we look at will involve a bit more topology than we've encountered so far, so first let's explore another topological concept that's useful in graph analytics and in understanding simplicial complexes.

Introduction to Homology

The basic topological premise of our next set of tools involves counting different dimensions of holes in an object or dataset. Consider a piece of paper with a hole in the middle of it or a basketball with a sphere of air inside it. These are holes of different dimensions, and each hole separates connected pieces of an object from other pieces of itself. When these holes exist in manifolds or functions, we can systematically study them and classify objects or spaces based on the number and dimension of these holes.

Homology is the counting of varying-dimensional holes (connected components, circles, spheres, voids, and so on) within a given object or space, usually to classify that object or space. For low-dimensional spaces, this is fairly straightforward; you can actually build a physical model of the space and count the holes. However, there are also variants of homology that allow topologists to distinguish between different types of objects and spaces that may be higher dimensional or strangely shaped without requiring a physical model.

Numbers corresponding to holes in each dimension create a handy collection of values, called *Betti numbers*, that organize the number and type of hole within a given object or space such that each object can be classified and studied alongside other objects whose numbers match. If you're familiar with algebraic topology, this is a standard procedure for the classification of abstract mathematical structures. Commonly, these numbers are stored in a vector. It's a bit abstract, but we'll go through some simple examples.

Examples of Betti Numbers

Many sports involve using a ball, but not all balls are the same, topologically speaking. Basketballs and baseballs are both round balls in three-dimensional space. Basketballs are usually bigger than baseballs, but if there were a child's toy basketball of the same size as a baseball, one might look at them and think they are quite similar.

Figure 4-5: An example baseball and basketball, which look similar but are topologically distinct

Topologically, though, they are quite distinct. These two balls differ in second Betti numbers, which count three-dimensional voids in an object. A vector of Betti numbers is an infinite sequence of numbers representing the number of holes in each dimension, starting with connected components on the zeroth number position and moving to circles (first number position), voids (second number position), and higher-dimensional voids (starting from the third position and going to infinite position). In practice, most datasets don't have many holes past the first Betti number, so we can fill the rest of the vector with zeros. The hollow basketball has a hole past the first Betti number because it contains a void, giving a vector of Betti numbers (1, 0, 1, 0, . . .), while the solid baseball has no holes of any dimension, corresponding to a Betti number vector of (1, 0, 0, 0, . . .).

Some objects have more than one hole in a given dimension. For instance, imagine gluing a second basketball to the outer surface of the basketball in Figure 4-5. This object would obviously have another void, yielding a Betti number vector of (1, 0, 2, 0, . . .). A donut, or *torus*, has a vector of (1, 2, 1, 0, . . .), as it has two open circles defining the ends of the tube, which form a void when connected at the ends. Figure 4-6 shows the classical construction of a torus from a sheet of paper.

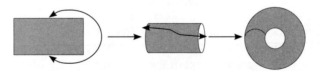

Figure 4-6: The construction of a torus from a sheet of paper connected at the edges

It's fairly easy to classify objects and spaces that can be easily visualized in three dimensions. However, many datasets used in the industry involve more than three dimensions, and comparisons and classifications of these objects require algorithms that can discern the Betti numbers associated with those objects; among these are genomics datasets (which can involve million-dimensional spaces), video sequences, and multivariate time series.

The Euler Characteristic

One of the topology-based metrics shows up both in the analysis of networks and in their higher-dimensional simplicial complex cousins, and it ties back to the notion of curvature introduced in prior chapters. The *Euler characteristic*, often given the notation of χ, provides a single number to summarize a topological space and is a topological invariant, meaning that the topological quantity being calculated does not change as the space is continuously deformed (stretched, twisted, or otherwise manipulated without tearing the space). The Euler characteristic can be defined using Betti numbers; technically, computing the Euler characteristic this way involves an alternating sum of Betti numbers (zeroth Betti number − first Betti number + second Betti number − third Betti number + fourth Betti number . . . up until the highest Betti number that exists).

The Euler characteristic can also be defined through the dimensions of the simplicial complex (number of vertices − number of edges + number of triangles − number of mutual 4-way interactions + . . .). However, vertices included in an edge aren't counted in the number of vertices. A triangle that makes up part of a mutual four-way interaction won't be counted either.

However, there is an easy way to obtain the largest pieces of a network or its higher-dimensional simplicial complex using an igraph function related to maximal cliques (as mentioned earlier). Maximal k − cliques denote and count the $k − 1$ simplices of the full simplicial complex derived from the network. They're a convenient way to build the full simplicial complex and keep track of the pieces involved at each n-way interaction. Let's add to the script in Listing 4-2 to count the maximal cliques in the author's network:

```
#create a table counting the number of k+1 simplices in the simplicial complex
summ<-as.numeric(summary(cl)[,1])
jjj<-table(summ)
```

This code creates a table summarizing the maximal cliques in the network that we previously computed. The result should yield 11 two-cliques (one-simplices, or edges), 6 three-cliques (two-simplices, or triangles), and 1 four-clique (three-simplices, or a mutual four-way interaction). We can plug these values into the Euler characteristic formula:

$$\chi = 0 \text{ vertices} - 11 \text{ edges} + 6 \text{ triangles} - 1 \text{ tetrahedron}$$

This gives a χ of −6. Recent studies have shown that most real-world networks have negative Euler characteristics. There's a very interesting reason that network data tends toward negative Euler characteristics related to the curvature of the network. Negative curvature in graphs is associated with the robustness of the network; biological networks with highly negative curvature can often withstand loss of function within parts of the network without adverse effects on the organism. The *Gauss–Bonnet theorem* relates the Euler characteristic, defined through homology, and the curvature of the object, including the manifold's curvature and the curvature of the manifold's boundary. There have been some recent attempts to link network analytics tools such as homology and Forman–Ricci curvature for a deeper study into network properties. This is a deep result in a branch of mathematics called *differential geometry* that connects an object's local geometry to its global topology, and it's a newer area of study in network science. Now that we know network topology and geometry are related to each other, let's look at a topological tool called *persistent homology*.

Persistent Homology

One of the most common topology-based algorithms used in data analysis today is persistent homology, which has been applied in genomics, healthcare, economics, energy, psychometrics, and many other fields. In essence, the idea of the persistent homology algorithm is to build a point cloud from the data, filter it into a series of simplicial complexes based on different thresholds of the data (akin to an MRI), and track topological features, such as holes or voids, appearing and disappearing in each slice. For instance, consider the three slices of cheese in Figure 4-7, each containing holes in the shape of circles; these circles affect the first Betti numbers of the datasets.

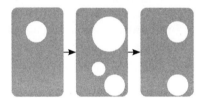

Figure 4-7: Three slices of a cheese block containing holes in different places

In Figure 4-7 one hole appears in all three slices, another appears in only the middle slice, and one appears in two slices. Holes and voids can be of different sizes in real data, and as we move across slices, holes might grow or shrink in diameter. Persistent homology algorithms have thresholds for both the lifetime of a feature and the minimum size considered for measuring a hole. In our example, we have features that are likely noise (either too small a radius or only appearing in one slice of our cheese) and features that are likely real features in the dataset (such as the void appearing in all three slices). Let's unpack this intuition.

Say we want to compare two datasets to see whether they are collected from the same distribution or shape. This is common when matching image data. While image data rarely comes with cheese holes, circles come up in image data quite frequently in the form of eyes.

Technically speaking, by varying the distances used to build the simplicial complex from the point cloud data (or filtering), you can track various Betti numbers through the filtration and assign each hole in the data an importance score, with important features lasting over longer filtration distances (longer *persistence*, in the parlance of persistent homology). In Figure 4-7, the hole that appears in all three slices would be considered the most important feature, and the hole that appears in only the second slice might be a result of noise in the data. These features can then be plotted on a *barcode* or *persistence diagram* that tracks these features' lifetimes (distance scale over which they exist in the filtration). We'll explore barcodes and persistence diagrams in the following example analysis.

In practice, datasets are usually examined only for low-dimensional holes and features due to computational issues, and the zeroth (connected components) and first (circles) Betti numbers are used most commonly unless you are explicitly computing high-dimensional shape data. The example in Figure 4-7 is connected in all three slices, so it has a zeroth Betti number of 1 across all slices. However, circles appear and disappear through the filtration, giving a barcode that looks like Figure 4-8.

Figure 4-8: A diagram plotting the persistence of features (holes) captured in the box of Figure 4-7

The barcode shows the time at which features appear and disappear. For instance, in Figure 4-8, we can see a feature that appears at time 2 and disappears at time 3 (our bottom cheese hole in Figure 4-7). The sequence of connected components across the data slices has a curious relationship with another machine learning method, single-linkage hierarchical clustering, in which clusters at each height level correspond to the connected components at that particular slice. When both techniques use the same distance metric, the results are actually identical; however, the persistent homology approach will give more information than single-linkage hierarchical clustering's dendrogram regarding the structure of the data. This means that machine learning practitioners can choose the technique that fits the problem best, as these two options come with their own plots and statistical tests. For instance, with a nontechnical audience, single-linkage hierarchical clustering might be preferable, as dendrograms and heatmaps are more familiar to biologists or social scientists.

Comparison of Networks with Persistent Homology

Within the realm of network analytics, persistent homology can be a useful way to compare network structures to see if different networks have the same underlying geometry. Let's explore this further with an application to simulated networks. In neuroscience, it's common to translate fMRI or PET data into a network structure, where different regions of the brain are translated to vertices and connected to other regions of the brain based on activity patterns (sequential activation of an area, for instance, or co-activation of multiple regions during one task). Often, outcomes of interest involve comparing groups of patients, either healthy patients against a group of patients with a particular neurological or psychological disorder or two disease groups, to understand differences in the brain activation patterns across disorders.

We'll explore the use of persistent homology in the comparison of two such networks. Because fMRI data isn't readily available as open source, we'll simulate networks in igraph that are approximately the size of brain imaging networks; this will demonstrate how this methodology would be applied to imaging data that has been transformed to network data.

The igraph package allows you to simulate many types of network data, including Erdös–Renyi graphs, scale-free graphs, and Watts–Strogatz graphs. We'll create each of these types of graphs using the script in Listing 4-3.

```
#simulate three graphs using the igraph package for further comparison
library(igraph)

#create an Erdos-Renyi graph
g1<-erdos.renyi.game(30,0.3)

#create a scale-free graph
g2<-sample_pa(30,power=2.5,directed=F)

#create a Watts-Strogatz graph
g3<-sample_smallworld(2,5,3,0.3)

#plot the three graphs created
plot(g1,main="Erdos-Renyi Graph")
plot(g2,main="Scale-Free Graph")
plot(g3,main="Watts-Strogatz Graph")
```

Listing 4-3: A script that simulates three different types of network structures for statistical comparison

Listing 4-3 creates three different types of networks that can later be compared via persistent homology; it also visualizes the networks, which should yield something similar (but probably not identical) to Figure 4-9.

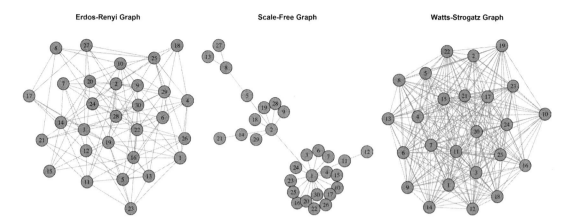

Figure 4-9: Plots of the three simulated network types

Figure 4-9 shows very different types of graphs. The scale-free graph in the middle includes a hub with many vertices connected to the hub but not to other vertices. The Erdös–Renyi graph on the left and the Watts–Strogatz graph on the right have many more interconnections, but the Watts–Strogatz model seems to have more structure connecting vertices into cliques, rather than randomly connecting vertices.

Let's apply persistent homology to these networks and compare the distance between persistence diagrams among these networks by adding the following to Listing 4-3; again, your results may vary given the simulation of each network type:

```
#load TDA package
library(TDAstats)

#get adjacency matrices
m1<-as.matrix(get.adjacency(g1))
m2<-as.matrix(get.adjacency(g2))
m3<-as.matrix(get.adjacency(g3))

#compute persistent homology
d1<-calculate_homology(m1,dim=2,format="cloud")
d2<-calculate_homology(m2,dim=2,format="cloud")
d3<-calculate_homology(m3,dim=2,format="cloud")

#plot persistence diagrams
plot_persist(d1)
plot_persist(d2)
plot_persist(d3)

#compute distances among graphs
w1<-phom.dist(d1,d2,limit.num=0)
w2<-phom.dist(d1,d3,limit.num=0)
w3<-phom.dist(d2,d3,limit.num=0)
```

This addition derives an adjacency matrix from each of the simulated graphs and computes a persistence diagram from this adjacency matrix, which is then compared through the distances between the zeroth homology groups. This script should produce three persistence diagrams that look like Figure 4-10 (note they won't be identical, as each run will produce something slightly different).

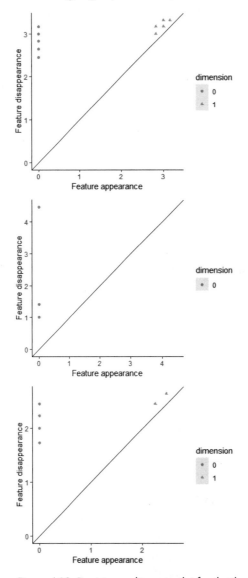

Figure 4-10: Persistence diagram plot for the three simulated network types (from top to bottom: Erdös–Renyi, scale-free, and Watts–Strogatz)

Figure 4-10 shows varying topological features found in each of the network types. The Watts–Strogatz network and Erdös–Renyi graphs both produce many large zeroth homology features (the dots), while the scale-free graph has a variety of zeroth homology feature sizes. The scale-free graph does not have higher-order homology features, while the other two graphs have first homology features (the triangles), albeit very near the diagonal line (suggesting that they may be noise). A point directly on the diagonal line is a feature that is in only one slice of the data; the farther from the diagonal line a point lies, the longer it has existed in the data. With respect to our three simulated networks, it's hard to tell if the scale-free and Watts–Strogatz graphs differ significantly from the Erdös–Renyi graph just by looking at the persistence diagrams.

We can add to our script to derive a null distribution for the Erdös–Renyi persistence diagram and use a special distance metric, Wasserstein distance, to statistically test the structural differences between the Erdös–Renyi persistence diagram and the scale-free and Watts–Strogatz persistence diagrams:

```
#get Wasserstein distance between random graphs with the same structure
ww<-rep(NA,100)

for (i in 1:100){
  g1<-erdos.renyi.game(30,0.3)
  g2<-erdos.renyi.game(30,0.3)
  m1<-as.matrix(get.adjacency(g1))
  m2<-as.matrix(get.adjacency(g2))
  d1<-calculate_homology(m1,dim=2,format="cloud")
  d2<-calculate_homology(m2,dim=2,format="cloud")
  ww[i]<-phom.dist(d1,d2,limit.num=0)
}

#compute 95% confidence intervals from the simulated null distribution
quantile(ww,c(0.025,0.975))
```

This script creates a null distribution of Erdös–Renyi persistence diagrams from the same distribution that the original persistence diagram was constructed from; your results may vary, given the random component to the simulation piece. Quantiles of our null distribution give a confidence interval of (0.91, 8.36), which includes quite a bit smaller distances than the distances computed between the persistence diagrams of the Erdös–Renyi graph and the Watts–Strogatz graph (23.59) and between the persistence diagrams of the Erdös–Renyi graph and the scale-free graph (39.78). Thus, we can conclude that the structures of the Watts–Strogatz graph and the scale-free graph are not random. There is a significant structural component to each of these graphs.

This type of simulation can be very useful in testing differences between persistence diagrams of brain networks derived from fMRI and PET imaging studies, and it's easy to implement in R. This methodology can also be applied to other networks with a hypothesized underlying structure, such as social networks or power grids. Many other types of network analysis tools can also be used to compare graph structures, such as local and global metrics (including graph radius and diameter, degree distributions, clustering graph coefficients, and so on), and many of these comparisons haven't been explored much yet.

Summary

In this chapter, we filtered weighted networks to understand how network metrics change as edges are removed based on their weights. Then, we built simplicial complexes from network data to leverage several topological tools, including an extension of the degree metric, the Euler characteristic, and a filtration-based algorithm called persistent homology that can be used to compare networks. In the next chapter, we'll transition from network science to distance geometry as we explore how different measurement choices impact supervised and unsupervised learning algorithms.

5

GEOMETRY IN DATA SCIENCE

In this chapter, we'll explore several tools from geometry: we'll look at distance metrics and their use in *k*-nearest neighbor algorithms; we'll discuss manifold learning algorithms that map high-dimensional data to potentially curved lower-dimensional manifolds; and we'll see how to apply fractal geometry to stock market data. The motivation for this chapter follows, among other things, from the *manifold hypothesis*, which posits that real-world data often has a natural dimensionality lower than the dimensionality of the dataset collected. In other words, a dataset that has 20 variables (that is, a dimensionality of 20) might have a better representation in a 12-dimensional space or an 8-dimensional space. Given the curse of dimensionality, representing data in lower-dimensional spaces is ideal (particularly when the original dimensionality of a dataset is large, as in genomics or proteomics data). Choosing the right distance measurements needed to create these representations has important implications for solution quality.

Introduction to Distance Metrics in Data

Many machine learning algorithms depend on distance metrics, which provide a measure between points or objects in a space or manifold. Changes in choice of distance metric can impact machine learning performance dramatically, as we'll see later in this chapter. *Distance metrics* provide a measure between points or objects in a space or manifold. This can be relatively straightforward like using a ruler to measure the distance between two points on a flat sheet of paper, as demonstrated in Figure 5-1.

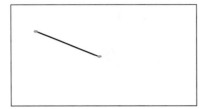

Figure 5-1: A plot of two points on a sheet of paper and the line connecting them

However, measuring the distance between two points on a sphere using a ruler will surely be a bit more complicated.

If you used a piece of string to limn out the shortest path connecting the two points on the sphere, as in Figure 5-2, you could mark the distance on the string and then use a ruler to measure that distance on the straightened-out string. This is akin to what is done with distances on manifolds, where *geodesics* (shortest paths between two points relative to the curved manifold) are lifted into the *tangent space* (a zero-curvature space defined by tangent lines, tangent planes, and higher-dimensional tangents) to measure distances.

Figure 5-2: A plot of two points on a sphere, along with the geodesic connecting them

We'll explore tangent spaces and their applications in machine learning in more depth in Chapter 6, but for now, you can think of lifting the string to a large sheet of paper and measuring its length with a ruler to measure distance outside of the curved space, where it's more difficult to establish a standard measurement. While geodesics and tangent spaces look counterintuitive, they follow from our knowledge of tangents in Euclidean geometry and derivatives in calculus.

However, there are other situations in which distances between two points are a bit more complicated. Consider walking from one house to another in the neighborhood, as shown in Figure 5-3.

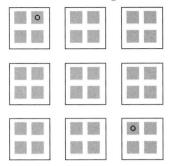

Figure 5-3: A plot of houses in a neighborhood, where one is walking between two houses

Unless one is able to walk through neighboring houses without running into exterior and interior walls (not to mention disgruntled neighbors!), it's not possible to draw a straight line or geodesic between the houses that gives a direct route, as you can see in Figure 5-4.

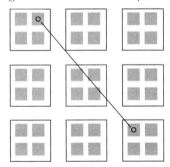

Figure 5-4: A plot of houses in a neighborhood, where one attempts a straight line between houses

Instead, it's a lot more practical to take the sidewalks (Figure 5-5).

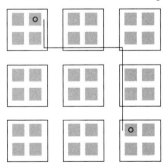

Figure 5-5: A plot of houses in a neighborhood, where one walks on the sidewalks between houses

Distance is often discrete, rather than continuous, or lies on a manifold with curvature. Understanding the geometry of the data space in which the data points live can give a good indication of what distance metric is appropriate for the data. In the following section, we'll go over some common distance metrics in machine learning, and then, in the sections after that, we'll apply these distances to k-NN algorithms and dimensionality reduction algorithms.

Common Distance Metrics

Given the nuances of measuring distance, it's important to understand some of the more common distance metrics used in machine learning, including one we briefly encountered in Chapter 4 (Wasserstein distance, used to compare persistent homology results). There are an infinite number of distance metrics, and some distance metrics have parameters that can give rise to an infinite number of variations. Thus, we cannot cover all possible distance metrics one could encounter in machine learning. We've left out some that are useful in recommender systems, such as cosine distance, as they are uncommon metrics within topological data analysis or network analysis applications. We'll explore some of the more common ones; if you're interested in going further, we suggest you explore the field of *metric geometry*.

Simulating a Small Dataset

Before we start exploring common distance metrics, let's simulate some data with Listing 5-1.

```
#create data
a<-rbinom(5,4,0.2)
b<-rbinom(5,1,0.5)
c<-rbinom(5,2,0.1)
mydata<-as.data.frame(cbind(a,b,c))

#create plot
library(scatterplot3d)
scatterplot3d(a,b,c,main="Scatterplot of 3-Dimensional Data")
```

Listing 5-1: *A script that simulates and plots a small dataset*

This script creates a dataset with three variables and plots points in a three-dimensional space. This should give a plot with points lying on the three axes (Figure 5-6).

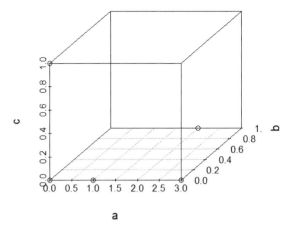

Figure 5-6: A plot of five points, all lying on axes defined by variables a, b, and c

This dataset includes the points shown in Listing 5-2, which we will use to calculate distances between points.

```
> mydata
  a b c
1 2 1 0
2 0 0 1
3 1 0 0
4 0 0 0
5 3 0 0
```

Listing 5-2: A matrix of the five points in the simulated dataset with random variables a, b, and c

Now that we have a dataset generated, let's look at some standard distance metrics that can be used to measure the distance between pairs of points in the dataset. R comes with a handy package, called the *stats* package (which comes with the base R installation), for calculating some of the common distance metrics used on data through the dist() function.

Using Norm-Based Distance Metrics

The first distances we'll consider are related. The *norm* of a function or vector is a measurement of the "length" of that function or vector. The norm involves summing distance differences to a power and then applying that power's root to the result. For the *Euclidean distance* between points, for example, the squares of differences are summed before taking the square root of the result. For single vectors (that is, a single data point), the norm will be a weighted distance from the origin, where the axes mutually intersect. You can think of this as the length of a straight line from the origin to the point being measured. Going back to the scatterplot of our points, this might be drawn like in Figure 5-7.

Scatterplot of 3-Dimensional Data

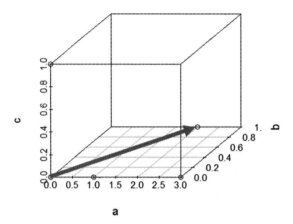

Figure 5-7: A plot of the five points with a straight line pointing to one of the points in the set

The most common norm used to measure metric distance between points is probably the Euclidean distance mentioned earlier, given by the L^2-norm, defined as the square root of squared distance between points where L is a placeholder for the vector (or vectors) and the exponent is the power of the norm (here, 2). This is the distance typically taught in high school geometry classes, and it is also referred to as the *Pythagorean distance*. We saw it in Figure 5-4, which showed a straight line of shortest distance between the houses (traveling as the bird flies above the houses). Statisticians typically use the square of the Euclidean distance metric when calculating squared errors in regression algorithms; for reasons we won't delve into here, using the square of Euclidean distance is very natural.

Related to the L^2-norm is the L^1-norm, or *Manhattan distance*. Manhattan distance calculations are much like the neighborhood example given in Figure 5-5. Manhattan distance is defined as the sum of point differences along each axis, with the axes' point differences summed into a final tally of axis distances. Let's say we have two points (0, 1) and (1, 0), which might represent whether a patient has a gene mutation in either gene of interest within a disease model. The Manhattan distance is (0 + 1) + (1 + 0), or the sum of point differences across all vector axes. In this example, we find the Manhattan distance is 2.

This metric is useful when working with count data or other discrete data formats, such as the example dataset generated earlier in this section. Figure 5-5 demonstrates this type of distance calculation, where the person needs to walk on the streets along the north-south and east-west axes.

Manhattan distance and L^1-norms often come up in applications of Lasso and elastic net regression, where it is used to set beta coefficients to 0 if they are within a certain distance of the origin, thereby performing variable selection and creating a sparse model. This is useful in situations where the dimensionality of the independent variable set is high (such as in genomic data).

A generalization of both the L^1-norm and L^2-norm is the *Minkowski distance*, which generalizes norms from L^3-norms to L^∞-norms. The L^∞-norm is another special instance of norm-based distances, dubbed the *Chebyshev distance*. Mathematically, Chebyshev distance is the maximum distance between points along any axis. It is often used in problems involving planned movements of machinery or autonomous systems (like drones).

As the dimension of the norm increases, the Minkowski distance values typically decrease and stabilize. Thus, Minkowski distances with high-dimensional norms can act as distance smoothers that rein in strange or unusually large distance calculations found with the Manhattan or Euclidean distance calculations. Minkowski distance does impose a few conditions, including that the zero vector has a length of zero, that application of a positive scalar multiple to a vector does not change the vector's direction, and that the shortest distance between two points is a straight line (known as the *triangle inequality condition*). In the dist() function of R, the dimension of the norm is given by the parameter p, with p=1 corresponding to Manhattan distance, p=2 corresponding to Euclidean distance, and so on.

A special extension of Manhattan distance is the *Canberra distance*, which is a weighted sum of the L^1-norm. Technically, Canberra distance is computed by finding the absolute value of the distance between a pair of points divided that by the sum of the pair of points' absolute values, which then is summed across point pairs. It can be a useful distance metric when dealing with outliers, intrusion detection, or mixed types of predictors (continuous and discrete measures). The example point in Figure 5-7 likely isn't a statistical outlier, but it certainly lies in a different part of the data space than the other simulated points.

Let's run these distances and compare the results on the dataset we simulated earlier in this section; add the following to the code in Listing 5-1:

```
#run distance metrics on example dataset
d1<-dist(mydata,"euclidean",upper=T,diag=T)
d2<-dist(mydata,"manhattan",upper=T,diag=T)
d3<-dist(mydata,"canberra",upper=T,diag=T)
d4<-dist(mydata,"minkowski",p=1,upper=T,diag=T)
d5<-dist(mydata,"minkowski",p=2,upper=T,diag=T)
d6<-dist(mydata,"minkowski",p=10,upper=T,diag=T)
```

This code computes distance metrics for Euclidean, Manhattan, Canberra, and Minkowski distances applied to our example dataset. Looking at Euclidean distance measurements between pairs of points in the simulated dataset, shown in Table 5-1, we see values with many decimal points for many pairs of points, owing to the square root involved in calculating the Euclidean distance.

Table 5-1: Euclidean Distance Calculations Between Pairs of Points in Listing 5-2's Matrix

Euclidean	1	2	3	4	5
1	0	2.44949	1.414214	2.236068	1.414214
2	2.44949	0	1.414214	1	3.162278
3	1.414214	1.414214	0	1	2
4	2.236068	1	1	0	3
5	1.414214	3.162278	2	3	0

Moving on to Manhattan distance (Table 5-2), the distances between pairs of points become whole numbers, as the calculation involves discrete steps along each axis separating the points.

Table 5-2: Manhattan Distance Calculations Between Pairs of Points in Listing 5-2's Matrix

Manhattan	1	2	3	4	5
1	0	4	2	3	2
2	4	0	2	1	4
3	2	2	0	1	2
4	3	1	1	0	3
5	2	4	2	3	0

As expected, the Minkowski distance calculations match the Manhattan distance for p=1 and Euclidean distance for p=2. In Table 5-3, you can see Minkowski distances with p=1.

Table 5-3: Minkowski Distance Calculations Between Pairs of Points in Listing 5-2's Matrix

Minkowski p=1	1	2	3	4	5
1	0	4	2	3	2
2	4	0	2	1	4
3	2	2	0	1	2
4	3	1	1	0	3
5	2	4	2	3	0

The Canberra distance gives some similar and overlapping values with Manhattan distance. However, some distances are different (particularly pairs involving points 2 or 3), owing to the weighted parts of the distance calculation, as shown in Table 5-4.

Table 5-4: Canberra Distance Calculations Between Pairs of Points in Listing 5-2's Matrix

Canberra	1	2	3	4	5
1	0	3	2	3	1.8
2	3	0	3	3	3
3	2	3	0	3	1.5
4	3	3	3	0	3
5	1.8	3	1.5	3	0

For some points in Listing 5-2's distance matrix calculations, these three distances give the same distance score for a pair of points (such as for points 4 and 5). However, some of the distances are quite different when we increase the value of p (such as points 1 and 2). If we're using the distance metrics in a support vector machine classifier, we might end up with a very different line cutting our data into groups—or very different error rates.

There are other ways to modify or extend norm-based distances. One popular modification is like the Canberra distance: *Mahalanobis distance* applies a weighting scheme to Euclidean distance calculations before taking the square root of the result, such that Euclidean distance is weighted by the covariance matrix. If the covariance matrix is simply the identity matrix, Mahalanobis distance will collapse to Euclidean distance. If the covariance matrix is diagonal, the result is a standardized Euclidean distance. Thus, Mahalanobis distance provides a type of "centered" distance metric that can identify leverage points and outliers within a data sample. It's often used in clustering and discriminant analyses, as outliers and leverage points can skew results.

There's a simple way to calculate Mahalanobis distance in R: the mahalanobis() function. Let's add to our script again:

```
#run Mahalanobis distance metrics
#first use the covariance to center the data
d7<-mahalanobis(mydata,center=F,cov=cov(mydata))

#then center to one of the points of the data, in this case point 1
d8<-mahalanobis(mydata,center=c(2,1,0),cov=cov(mydata))

#then use the column means to center the data
d9<-mahalanobis(mydata,center=colMeans(mydata),cov=cov(mydata))
```

This code will calculate Mahalanobis distance with various centering strategies, yielding three different measures of leverage/weighted standard distance from a defined reference, detailed in Table 5-5.

Table 5-5: Mahalanobis Distance Results for the Individual Points from Figure 5-7's Matrix

Mahalanobis	1	2	3	4	5
Covariance only	6.857143	6.857143	0.857143	0	7.714286
Point 1	0	8	5.428571	6.857143	7.714286
Column means	3.2	3.2	0.628571	2.057143	2.914286

By using each point as a center, you can complete a full distance matrix similar to how the dist() function creates the distance matrix. You would simply loop through the individual points and append rows to a data frame.

A few interesting observations come out of the Mahalanobis distance calculations. When only covariance is used, the origin becomes the reference point for calculating distances, and point 4, which is located at the origin, has a Mahalanobis distance of 0. However, when column means are used to center the data, this point jumps to a much farther away value. This suggests that point 4 is quite far away from the column means, though it is perfectly centered at the origin. Another interesting trend involves point 3, which is quite close to both the origin and the centered column means, which come out to (1.2, 0.2, 0.2) in this dataset. Point 3 is located at (1, 0, 0), which is both near the origin and near this centered column mean. The other points are relatively far from both the origin and the centered column means.

We can add these column means to our plot of this data and visualize a bit of how Mahalanobis distance works by adding to our script again:

```
#add point to dataset created earlier in this section
colmean<-c(1.2,0.2,0.2)
mydata<-rbind(mydata,colmean)

#create plot
library(scatterplot3d)
scatterplot3d(mydata[,1],mydata[,2],mydata[,3],
main="Scatterplot of 3-Dimensional Data")
```

This code adds the column mean point to the original dataset to examine where the "middle" of the data should be located in the three-dimensional space; the code should yield a plot similar Figure 5-8.

Examining Figure 5-8 and comparing it to Figure 5-6, we can see that a point has been placed off the axes that does seem to occupy a central location among the five points. Finding the central location of a dataset helps in several data science tasks, including finding stealth outliers (outliers without extreme values for any one variable but far from most points in the multivariate dataset) and calculating multivariate statistics. Some points are closer to this central location than others, as our Mahalanobis results suggest; these are points 3 and 4 in our dataset, which are relatively close to the origin.

Scatterplot of 3-Dimensional Data

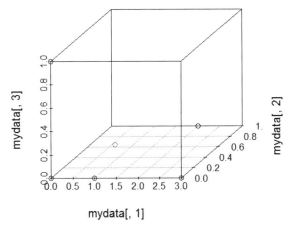

Figure 5-8: Mahalanobis distance with centering at column means for the individual points from Listing 5-2's matrix plotted with the column mean point shown off the axes

These distance differences come up in many machine learning applications, and we'll see how distance impacts machine learning performance and results when we apply these distances within a *k*-nearest neighbors problem later in this chapter. Performance can vary dramatically with a different choice of metric, and using the wrong metric can mislead model results and interpretation.

Comparing Diagrams, Shapes, and Probability Distributions

Norm-based distance metrics are not the only class of metrics possible in machine learning. As we saw in Chapter 4, it's possible to calculate distance between objects other than points, such as persistence diagrams. Roughly speaking, these types of metrics measure differences in probability distributions. We've already briefly used one of these, the Wasserstein distance, to compare persistence diagram distributions. Let's take a closer look now.

Wasserstein Distance

In loose terms, *Wasserstein distance* compares the piles of probability weights stacked in two distributions. It's often dubbed "earth-mover distance," as the Wasserstein distance measures the cost and effort needed to move probability piles of one distribution to turn it into the comparison distribution. For more mathematically sophisticated readers, the pth Wasserstein distance can be calculated by taking the expected value of the joint distribution marginals to the pth power, finding the infimum over all join probability distributions of those random variables, and taking the pth root of the result. However, the details of this are beyond what is expected of readers, and we'll stick with the intuition of earth-mover distance as we explore this metric. Let's visualize two distributions of dirt piles to build some intuition behind this metric (Figure 5-9).

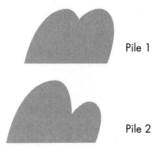

Figure 5-9: Two distributions of dirt piles akin to the type of probability density functions that could be compared using Wasserstein distance metrics

Pile 1 of Figure 5-9 has a large stack of dirt toward the left side that would need to be shoveled to the right piles if we were to transform Pile 1's distribution of dirt to Pile 2's. Our dirt mover will have to move quite a bit of dirt to transform Pile 1 into Pile 2. However, if our Pile 2 had a distribution of dirt piles closer to that of Pile 1, as in Figure 5-10, the amount of work to turn Pile 1 into Pile 2 would be less for our dirt mover.

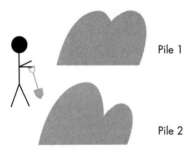

Figure 5-10: Two distributions of dirt piles akin to the type of probability density functions that could be compared using Wasserstein distance metrics, which measure the amount of work needed to be done by our dirt mover to transform Pile 1 into Pile 2

To get from Distribution 1 to Distribution 2, you can think of someone shoveling the dirt. The amount of dirt shoveled corresponds to the Wasserstein distance. Probability density functions that are very similar will have a smaller Wasserstein distance; those that are very dissimilar will have a larger Wasserstein distance.

Thus, Wasserstein distance can be a nice metric to use in comparing probability distributions—comparing theoretical distributions to sample distributions to see if they match, comparing multiple sample distributions from the same or different populations to see if they match, or even understanding if it's possible to use a simpler probability distribution in a machine learning function that approximates the underlying data such that calculations within a machine learning algorithm will be easier to compute.

Although we won't go into it here, some distributions—including count data, yes/no data, and even machine learning output structures like dendrograms—can be compared through other metrics. For now, let's look at another way to compare probability distributions, this time with discrete distributions.

Entropy

One common class of distance metrics involves a property dubbed *entropy*. Information entropy, defined by the negative logarithm of the probability density function, measures the amount of nonrandomness at each point of the probability density function. By understanding how much information is contained in a distribution at each value, it's possible to compare differences in information across distributions. This can be a handy tool for comparing complicated probability functions or output from machine learning algorithms, as well as deriving nonparametric statistical tests.

Binomial distributions come up often in data science. We might think that a random customer has no preference between two new interface designs (50 percent preferring A and 50 percent preferring B in an A/B test). We could estimate the chance that 10 or 20 or 10,000 customers prefer A over B and compare it to samples of actual customers providing us feedback. One assumption might be that we have different customer populations, including one that is very small. Of course, in practice, we don't know the actual preference distributions of our customer populations and may not have enough data to compare the distributions mathematically via a proportions test. Leveraging a metric can help us derive a test.

To understand this a bit more intuitively, let's simulate two binomial probability distributions with the code in Listing 5-3.

```
#create samples from two different binomial probability distributions
a<-rbinom(1000,5,0.1)
b<-rbinom(1000,5,0.4)

#create plot of probability density
plot(density(b),ylim=c(0,2),main="Comparison of Probability Distributions")
lines(density(a),col="blue")
```

Listing 5-3: A script that simulates different binomial probability distributions

Figure 5-11 shows that these binomial distributions have very different density functions, with information stored in different parts of the distribution.

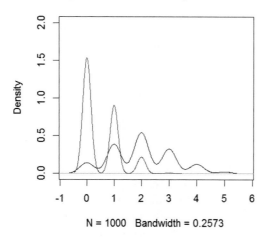

Figure 5-11: Two binomial distributions' density functions plotted for comparison

The black curve distribution, distribution *b*, includes a wider spread of information over more values than distribution *a*, which concentrates its information nearer to zero (lighter gray curve). Entropy-based metrics can be used to quantify this difference in information storage between distributions *a* and *b*. R provides many tools and packages for quantifying and comparing information entropy. Let's explore a bit further.

The philentropy package in R contains 46 different metrics for comparing probability distributions, including many metrics based on entropy. One of the more popular entropy metrics is the *Kullback–Leibler divergence*, which measures the relative entropy of two probability distributions. Technically speaking, the Kullback–Leibler divergence measures the expectation (sum for discrete distributions or integral for continuous distributions) of the logarithmic differences between two probability distributions. As such, it's a measurement of information gain or loss. This allows us to convert information entropy into a distribution comparison tool, which is useful when we're trying to compare differences between unknown or complicated probability distributions that might not be amenable to the usual statistical tools.

Let's develop our intuition by returning to our two binomial distributions, *a* and *b*. We'll calculate the Kullback–Leibler divergence between the two distributions by adding the following to Listing 5-3:

```
#load package
library(philentropy)

#calculate Kullback-Leibler divergence
kullback_leibler_distance(P=a,Q=b,testNA=T,unit="log2",epsilon=1e-05)
```

This code calculates the Kullback–Leibler divergence for the two binomial distribution samples generated in Listing 5-3. In this set of simulated data, the Kullback–Leibler divergence is 398.5428; another simulation of these distributions might yield a different divergence measurement. However, using nonparametric tests, we can compare this divergence value with the random error component of one of our distributions to see whether there is a statistically significant difference of entropy between distributions *a* and *b*. We can add to our script to create a nonparametric statistical test using entropy differences:

```
#create a nonparametric test
#create a vector to hold results from the simulation loop
test<-rep(NA,1000)

#loop to draw from one of the binomial distributions to generate
#a null distribution for one of our samples
for (i in 1:1000){
  new<-rbinom(1000,5,0.1)
  test[i]<-kullback_leibler_distance(P=a,Q=new,testNA=T,unit="log10",epsilon=1e-05)
}

#obtain the cut-off score for 95% confidence intervals, corresponding
#to values above/below which a sample would be considered statistically
#different than the null distribution
quantile(test,c(0.025,0.0975))
```

The confidence intervals from this test suggest confidence intervals of 1,427–1,475, which suggests that our distributions are significantly different. This is expected, as distributions *a* and *b* have very different values and ranges. Plotting the last distribution simulated (Figure 5-12) shows that the new distribution is a much better match to *a* than *b* is.

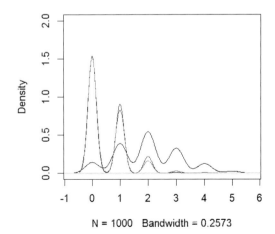

Figure 5-12: Three binomial distributions' density functions plotted for comparison, with two coming from samples of the same distribution

Using the Kullback–Leibler divergence, we've determined that a and b are different populations statistically. If we saw a confidence range including 104.2, we'd conclude that a and b likely come from the same population distribution. Statistical tests such as proportions tests exist to compare binomial distributions in practice, but some discrete distributions or sample sizes don't have easy statistical comparisons (such as comparisons of predicted class distributions coming out of two convolutional neural network classifiers).

Comparison of Shapes

As we saw in Chapter 4, sets of points and shapes, such as persistence diagrams, can be important data structures, and these objects can be measured and compared as well. The next three measures will deal with this situation in more depth. Let's start with an example of two circles with differing radii, as shown in Figure 5-13.

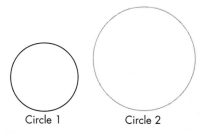

Figure 5-13: Two circles of differing radii

Now, let's imagine these two circles are paths in a park, and someone is walking their dog on a leash, with the dog following the outer path and the owner following the inner path. Figure 5-14 shows a visual.

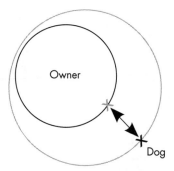

Figure 5-14: A dog and owner connected by a leash walking on different paths at a park

At some points, the owner and her dog are close together, and a small leash suffices to connect them. However, as they move counterclockwise, the distance between the owner and their dog increases, necessitating more leash to connect them; you can see this in Figure 5-15.

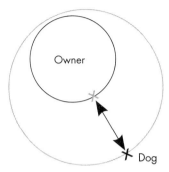

Figure 5-15: A dog and owner connected by a longer leash walking on different paths at a park

One historically important metric that compares distances between points on different shapes is *Fréchet distance.* The version of Fréchet distance that we'll consider here applies to discrete measurements (usually taken to be polygons). A grid graph is constructed from the polygons, and minmax paths are computed to find the maximum distance that two paths may be from each other. Many assumptions can be placed on the paths themselves and the synchronization of movement along those paths; the strictest requirements give rise to what's called a *homotopic* Fréchet distance, which has applications in many robotics problems. We'll return to homotopy applications in Chapter 8.

For now, in more lay terms, Fréchet distance is known as the dog-walking distance, and it has many uses in analytics. It can be used to measure not only the maximum distance between points on curves or shapes but the total distance between points on shapes or curves. Many R packages include functions to calculate one of the extant versions of Fréchet distance, including the TSdist time-series package, which is used in the following example. In this package, two time series are generated from ARMA(3, 2) distributions, with Series 3 containing 100 time points and Series 4 containing 120 time points. Time series are important in tracking disease progression in groups of patients, tracking stock market changes over time, and tracking buyer behavior over time.

Let's load these package-generated time series and plot them to visualize potential differences using the code in Listing 5-4.

```
#load package and time series contained in the TSdist package
library(TSdist)
data(example.series4)
data(example.series3)
my1<-example.series4
my2<-example.series3
```

```
#plot both time series
plot(my1,main="Time Series Plots of Series 4")
plot(my2,main="Time Series Plots of Series 3")
```

Listing 5-4: A script that loads and examines two time-series datasets

Figure 5-16 shows two time series with distinct highs and lows over time.

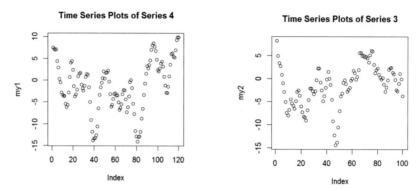

Figure 5-16: Plots of the two example time series

Notice the overlap between the time series isn't perfect, and we'd expect our comparisons to show some differences between these time series. With Fréchet distance, it's possible to measure both the maximum/minimum deviation (maximum/minimum leash length) between the time series and the sum of deviations across the full comparison set. We'll examine both of these for Series 3 and Series 4 by adding the following code to our script:

```
#calculate Frechet distance
dis1<-FrechetDistance(my1,my2,FrechetSumOrMax="sum")
dis2<-FrechetDistance(my1,my2,FrechetSumOrMax="min")
dis3<-FrechetDistance(my1,my2,FrechetSumOrMax="max")
```

This code calculates the minimum, maximum, and sum of Fréchet distances between the time series, which should yield a value of 402.0 for the sum of distances between the time-series curves, a value of 13.7 for the maximum distance between points on the time series curves, and a value of 0.03 for the minimum distance between points on the time series curves. This suggests that the time series have approximately the same values at some comparison points and values that are very different at other points. The sum of distances between the time-series curves will converge to the integral taken with continuous time across the series; this calculation can give a good tool for calculating areas between functions using discrete and fairly quick approximations.

There are other ways to compare shapes besides Fréchet distance, though, and these are sometimes preferable. Let's return to our two circles again and move them so that they are intersecting, such as in Figure 5-17.

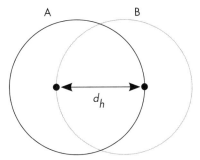

Figure 5-17: Plot of two intersecting circles

We can think of points on each circle, much like we did to measure Fréchet distance. Let's consider a point on circle B and its closest point on circle A (shown in Figure 5-17). Each point on circle B will have a closest point on circle A, and these form a collection of closest points. The points chosen in Figure 5-17 are a special set of points. They are the farthest apart of any points in our set of closest points. This means that the maximum we'd have to travel to hop from one circle to the other occurs with those two points. This distance is called the *Hausdorff distance*, and it is found in a lot of applications in early computer vision and image matching tasks. These days, it is mainly used for sequence matching, graph matching, and other discrete object matching tasks.

However, one of the limitations of Hausdorff distance is that the sets being compared must exist in the same metric spaces. Thus, while we can compare points on circles, we cannot directly compare points on a circle to those on a sphere with Hausdorff distance or points on a Euclidean plane to points on a positively curved sphere. The two objects being compared must be in the same metric space.

Fortunately, a solution to this conundrum exists. We can simply measure the farthest shortest distance from two metric spaces when those metric spaces are mapped to a single metric space while preserving the original distances between points within each space (called an *isometric embedding*). So, we could project the sphere and its points to tangent space to compare with other Euclidean spaces. Or we could embed two objects in a higher-dimensional space similar to what is done in kernel applications. This extension of Hausdorff distance is dubbed *Gromov–Hausdorff distance*.

Let's build some intuition around this metric. Say we have a triangle and a tetrahedron, as in Figure 5-18.

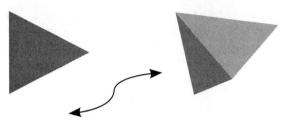

Figure 5-18: A triangle and tetrahedron, which exist in different-dimensional Euclidean spaces

One solution to this problem depicted in Figure 5-18 is to simply bring the triangle into three-dimensional Euclidean space and calculate the distances between the objects in three-dimensional Euclidean space. Perhaps part of the triangle overlaps with the tetrahedron when it is embedded in three-dimensional space, as shown in Figure 5-19.

Figure 5-19: A triangle and tetrahedron, both mapped into three-dimensional Euclidean space

It's now possible for us to compute the farthest point sets of the closest ones for these two objects, likely occurring at one of the far tips of the triangle or tetrahedron in this example.

R has a nice package for computing Gromov–Hausdorff distances (gromovlab), so we can easily implement this distance metric in R. Let's first simulate a small sample from a two-dimensional disc with the code in Listing 5-5.

```
#create two-dimensional disc sample
a<-runif(100,min=-1,max=1)
b<-runif(100,min=-1,max=1)

#create circle from uniform distribution and restrict to points within the
#circle
d<-a^2+b^2
```

```
w<-which(d>1)
mydata<-cbind(a,b)
mydata<-mydata[-w,-w]

#plot sample
plot(mydata,main="2-Dimensional Disc Sample",ylab="y",xlab="x")
```

Listing 5-5: *A script that samples from a two-dimensional disc to examine Gromov–Hausdorff distance*

This should give a rough disc shape when plotted; take a look at Figure 5-20.

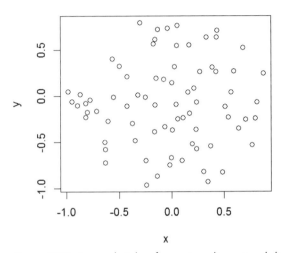

Figure 5-20: *A sample taken from a two-dimensional disc*

Now, let's add to Listing 5-5 to simulate a sample of the same size from the same uniform distribution that was used to generate our two-dimensional disc sample:

```
#create a uniform sample from a line segment
x<-sort(runif(dim(as.data.frame(mydata))[[1]],min=-1,max=1))
```

The code we've added samples one of the line segments to give a one-dimensional space. This gives us two Euclidean spaces of differing dimension. Now, we can compute the distances between points in each sample's native space (two-dimensional Euclidean space for the disc, one-dimensional Euclidean space for the line). From there, we can compute the Gromov–Hausdorff distance between our samples by adding the following code:

```
#load the package and calculate the distance matrices for use in calculations
library(gromovlab)
m1<-dist(as.matrix(mydata))
m2<-dist(as.matrix(x))
```

```
#calculate distance metric and compare distances with Gromov-Hausdorff
gromovdist(m1,m2,"lp",p=2)
```

This code allows us to compare distance matrices between points in the samples. For this random sample, the Gromov–Hausdorff distance is 5.8. We could simulate a nonparametric test based on our metric as we did in Listing 5-3 to help us determine if the embeddings of our disc and our line are the same statistically. Changing the metric parameters may change the significant differences between embeddings or the quality of an embedding, as we saw earlier in this chapter when we compared Canberra, Manhattan, and Euclidean distances. Interested readers are encouraged to play around with the embedding parameters, set up their own nonparametric tests, and see how the results vary for Gromov–Hausdorff distances for our disc and line sample.

The lp parameter allows one to use the norm-based metrics examined earlier in this chapter. For this particular comparison, we've used the Euclidean norm, as both samples lie in Euclidean spaces and the distance matrices ingested are defined by the Euclidean norm. Other norms, such as the Manhattan or Chebyshev, are possible and perhaps preferable for other problems, and the package is equipped to handle graphs and trees, as well as distance matrices. One thing to note about this particular package is that the algorithm searches through all possible isometric embeddings, so the compute time and memory needed may be large for some problems.

K-Nearest Neighbors with Metric Geometry

Metric geometry shows up in many algorithms, including *k*-nearest neighbor (*k*-NN) analyses, which classify observations based on the classifications of objects near them. One way to understand this method is to consider a high school cafeteria with different cliques of students, as shown in Figure 5-21.

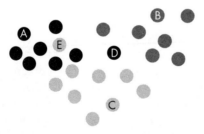

Figure 5-21: A high school cafeteria with three distinct student cliques

In the cafeteria shown in Figure 5-21, three student cliques exist: a dark gray clique, a gray clique, and a light gray clique. Students tend to stay near their group of friends, as exhibited by students A, B, and C. These students are surrounded by their friends, and classifying them using an arbitrary number of students standing nearest them according to a distance metric (like Euclidean distance or number of floor tiles between students) would give a pretty accurate classification into student cliques.

However, there are a few students, such as students D and E, who are located near other cliques or among all three cliques. Student E might be part of the popular clique (bottom center) and also part of the varsity athlete clique (top left), and student D might be popular, a varsity athlete, and a math team member (top right). Depending on how many students located near students D and E are considered in classifying them into a clique, they may belong to their main clique or be incorrectly reassigned to a new clique, of which they may fit but not consider their main clique. For instance, the closest 10 students may assign student E correctly to the popular group, while the closest 2 students would not.

Thus, k-NN methods rely on both a neighborhood size (in this instance, the number of students nearest the student of interest) and a distance metric defining which students are closest to the student of interest. Let's look at little more closely at how distance metric can impact k-NN classification accuracy with five nearest neighbors in a simulated dataset with three variables impacting classification and three noise variables, given in Listing 5-6, which uses the knnGarden package (and includes many of the distances covered in the simulated data analyzed with norm-based distance metrics earlier in this chapter). You'll first need to download the package (*https://cran.r-project.org/web/packages/knnGarden/index.html*) and install it locally.

```
#install package (and devtools if not installed)
#your local computer might save the .tar file in a different path than ours
library(devtools)
install_local("~/Downloads/knnGarden.tar")

#create data
a<-rbinom(500,4,0.2)
b<-rbinom(500,1,0.5)
c<-rbinom(500,2,0.1)
d<-rbinom(500,2,0.2)
e<-rbinom(500,1,0.3)
f<-rbinom(500,1,0.8)
class<-a+e-d-rbinom(500,2,0.3)
class[class>=0]<-1
class[class<0]<-0
mydata<-as.data.frame(cbind(a,b,c,d,e,f,class))

#partition data into training and test sets (60% train, 40% test)
s<-sample(1:500,300)
train<-mydata[s,]
test<-mydata[-s,]

#create KNN models with different distances and five nearest neighbors
library(knnGarden)
```

```
#Euclidean
ke<-knnVCN(TrnX=train[,-7],OrigTrnG=train[,7],TstX=test[,-7],
K=5,method="euclidean")
accke<-length(which(ke==test[,7]))/length(test[,7])

#Canberra
kc<-knnVCN(TrnX=train[,-7],OrigTrnG=train[,7],TstX=test[,-7],
K=5,method="canberra")
acckc<-length(which(kc==test[,7]))/length(test[,7])

#Manhattan
km<-knnVCN(TrnX=train[,-7],OrigTrnG=train[,7],TstX=test[,-7],
K=5,method="manhattan")
acckm<-length(which(km==test[,7]))/length(test[,7])
```

Listing 5-6: A script that generates and classifies a sample through k-NN classification with varying distance metric and five neighbors

Listing 5-6 creates a sample and then runs the *k*-NN algorithm to classify points based on different distance metrics, including Manhattan, Euclidean, and Canberra distances. In this particular simulation, all of our distances yield similar accuracies (Euclidean distance of 81 percent, Manhattan distance of 81 percent, and Canberra distance of 82 percent). We can consider a larger neighborhood by modifying Listing 5-6 to include 20 nearest neighbors.

```
#create KNN models with different distance metrics and 20 nearest neighbors

#Euclidean
ke<-knnVCN(TrnX=train[,-7],OrigTrnG=train[,7],TstX=test[,-7],
K=20,method="euclidean")
accke<-length(which(ke==test[,7]))/length(test[,7])

#Canberra
kc<-knnVCN(TrnX=train[,-7],OrigTrnG=train[,7],TstX=test[,-7],
K=20,method="canberra")
acckc<-length(which(kc==test[,7]))/length(test[,7])

#Manhattan
km<-knnVCN(TrnX=train[,-7],OrigTrnG=train[,7],TstX=test[,-7],
K=20,method="manhattan")
acckm<-length(which(km==test[,7]))/length(test[,7])
```

This script modifies the functions that calculate the *k*-NN model, with the changes marked in bold; we changed the parameter to K=20. With this particular simulated dataset, there are dramatic differences in classification accuracy when 20 nearest neighbors are considered. Euclidean and Manhattan distances give a slightly worse accuracy of 78.5 percent, and Canberra distance gives a much worse accuracy of 57 percent. Neighborhood size matters quite a bit in accuracy for Canberra distance, but it plays a lesser role for Euclidean and Manhattan distances. Generally speaking, using larger numbers of nearest neighbors smooths the data, similarly to how a weighted average might. These results suggest that, for our Canberra distance, adding a larger number of nearest neighbors might be smoothing the data too much. However, our Manhattan and Euclidean distance runs don't show this smoothing effect and retain their original high performance. As our example shows, the choice of distance metric can matter a lot in algorithm performance—or it can matter little. Distance metrics thus function like other parameter choices in algorithm design.

k-NN models are among the most closely tied to metric geometry and neighborhoods, though many other methods rely on distance metrics or neighborhood size. There are many recent papers that suggest a multiscale approach to algorithm neighborhood definition can improve algorithm performance, including applications in k-NN regression, deep learning image classification, and persistent graph and simplex algorithms (including persistent homology), and this nascent field has grown in recent years.

One branch of machine learning where the choice of distance metric matters a lot is in dimensionality reduction, where we're mapping a high-dimensional dataset to a lower-dimensional space. For instance, imagine we have a genomic dataset for a group of patients including 300 gene loci of interest. That's a bit too much to visualize for a stakeholder on a PowerPoint slide. However, if we find a good mapping to two-dimensional space, we can add a scatterplot of our data to the slide deck in a way that is much easier for humans to process.

Manifold Learning

Many dimensionality reduction algorithms also involve distance metrics and k-NN calculations. One of the most common dimensionality reduction algorithms, *principal component analysis (PCA)*, helps wrangle high-dimensional data into lower-dimensional spaces using a linear mapping between the original high-dimensional space and a lower-dimensional target space. Essentially, PCA finds the ideal set of linear bases to account for the most variance (packing in most of the relevant information related to our data) with the fewest linear bases possible; this allows us to drop many of the data space's bases that don't contain much relevant information. This helps us visualize data that lives in more than three dimensions; it also decorrelates predictors being fed into a model.

However, as noted, PCA assumes that data lives in a geometrically flat space and is mapped to a lower-dimensional flat space. As we've seen, this isn't always the case, and Euclidean metrics can give different distance results than other distance metrics. Recently, many attempts to relax different assumptions and generalize dimensionality reduction to manifolds have provided a new class of dimensionality reduction techniques, called *manifold learning*. Manifold learning allows for mappings to lower-dimensional spaces that might be curved and generalizations of PCA to include metrics other than Euclidean distance. A *manifold* is a space that is locally Euclidean, with Euclidean space being one example of a manifold, so some people refer to *manifold learning* as an umbrella to this more general framework.

Using Multidimensional Scaling

One of the older manifold learning algorithms is *multidimensional scaling (MDS)*. MDS considers embeddings of points into a Euclidean space such that distances between points, which can be Euclidean distances, are preserved as best as possible; this is done through the minimization of a user-defined cost function. Defining distances and cost functions via Euclidean distance yields the same results as PCA. However, there is no need to limit oneself to Euclidean distance with MDS, and many other metrics might be more suitable for a given problem. Let's explore this further with a small dataset and different distance matrices as input to our MDS algorithm; take a look at the code in Listing 5-7.

```
#create data
a<-rbinom(100,4,0.2)
b<-rbinom(100,1,0.5)
c<-rbinom(100,2,0.1)
d<-rbinom(100,2,0.2)
e<-rbinom(100,1,0.3)
f<-rbinom(100,1,0.8)
mydata<-as.data.frame(cbind(a,b,c,d,e,f))

#create distance matrices using different distance metrics
m1<-dist(mydata,upper=T,diag=T)
m2<-dist(mydata,"minkowski",p=10,upper=T,diag=T)
m3<-dist(mydata,"manhattan",upper=T,diag=T)
```

Listing 5-7: A script that generates an example dataset and calculates distance matrices from the dataset

Now that we have generated some data and have calculated three different distance metrics, let's see how the choice of distance metric impacts MDS embeddings. Let's compute the MDS embeddings and plot the results by adding to Listing 5-7.

```
#reduce dimensionality with MDS to two dimensions
c1<-cmdscale(m1,k=2)
c2<-cmdscale(m2,k=2)
c3<-cmdscale(m3,k=2)

#plot results
plot(c1,xlab="Coordinate 1",ylab="Coordinate 2",
main="Euclidean Distance MDS Results")
plot(c2,xlab="Coordinate 1",ylab="Coordinate 2",
main="Minkowski p=10 Distance MDS Results")
plot(c3,xlab="Coordinate 1",ylab="Coordinate 2",
main="Manhattan Distance MDS Results")
```

Our addition to Listing 5-7 should give plots that look different from each other. In this example, the plots (shown in Figure 5-22) do vary dramatically depending on the metric used, suggesting that different distance metrics result in embeddings to different spaces.

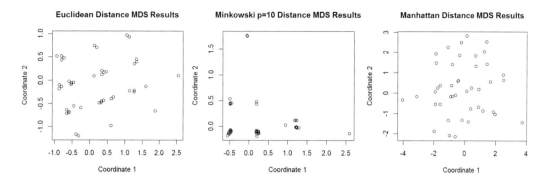

Figure 5-22: A side-by-side view of MDS results, which vary by distance metric chosen

The plot results in Figure 5-22 suggest that Minkowski distance yields quite different results than Euclidean or Manhattan distances; many points are bunched together in the Minkowski-type MDS result, which suggests it may not distinguish between pairs of points as well as the other metrics. However, the differences between Euclidean and Manhattan distance MDS results are less dramatic, with points spread out a lot more than in the case of our Minkowski distance.

Extending Multidimensional Scaling with Isomap

Some manifold learning algorithms extend MDS to other types of spaces and distance calculations. *Isomap* extends MDS by replacing the distance matrix with one of geodesic distances between points calculated from a neighborhood graph. This replacement of distance calculations with geodesic distances allows for the use of distances that naturally exist on spaces that are not flat, such as spheres (for instance, geographic information system data) or organs in a human body examined through MRIs. Most commonly, distances are estimated by examining a point's nearest neighbors. This gives Isomap a neighborhood flavor and a way to investigate the role of scaling through the variance of the nearest neighbor parameter.

Let's explore this modification by adding to Listing 5-7, which simulated a dataset and explored MDS. We'll use Euclidean distance as a dissimilarity measure, though other distance metrics can be used much as they were with MDS. To understand the role of neighborhood size, we'll create neighborhoods of 5, 10, and 20 nearest neighbors:

```
#create Isomap projections of the data generated in Listing 5-6
library(vegan)

i1<-scores(isomap(dist(mydata),ndim=2,k=5))
i2<-scores(isomap(dist(mydata),ndim=2,k=10))
i3<-scores(isomap(dist(mydata),ndim=2,k=20))

#plot results
plot(i1,xlab="Coordinate 1",ylab="Coordinate 2",main="K=5 Isomap Results")
```

```
plot(i2,xlab="Coordinate 1",ylab="Coordinate 2",main="K=10 Isomap Results")
plot(i3,xlab="Coordinate 1",ylab="Coordinate 2",main="K=20 Isomap Results")
```

This snippet of code applies Isomap to the generated dataset in Listing 5-7 using Euclidean distance. Other distance metrics can be used and may give different results, as shown in the MDS analyses. The results of the Isomap analyses suggest that neighborhood size doesn't play a large role in determining results for this dataset, as shown by the scales for each coordinate in the Figure 5-23 plots.

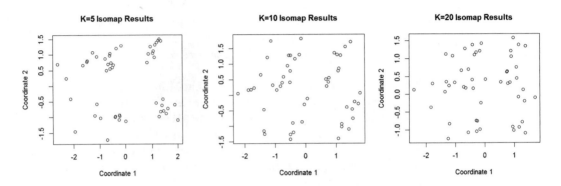

Figure 5-23: A side-by-side view of Isomap results, which vary by number of nearest neighbors

MDS and Isomap aim to preserve distance between points regardless of how far apart the points lie on the data manifold, resulting in global preservation of distance. Other global manifold learning algorithms, which preserve distances between points that are not in the same neighborhood, exist. If you're interested, you can explore global algorithms such as kernel PCA, autoencoders, and diffusion mapping.

Capturing Local Properties with Locally Linear Embedding

Sometimes global properties of the manifold aren't as important as local properties. In fact, from the classical definition of a manifold, local properties might sometimes be more interesting. For instance, when we're looking for nearest neighbors to a point, points that are very far way geometrically probably won't be nearest neighbors of that point, but points that are nearby could be nearest neighbors with information that needs to be preserved in a mapping between higher-dimensional and lower-dimensional spaces. Local manifold learning algorithms aim to preserve the local properties with less focus on preserving global properties in the mapping between spaces.

Locally linear embedding (LLE) is one such local manifold learning algorithm, and it is one of the more often used manifold learning algorithms. Roughly speaking, LLE starts with a nearest neighbor graph and then proceeds to create sets of weights for each point given its nearest neighbors. From there, the algorithm calculates the mapping according to a cost function and the preservation of the nearest neighbor weight sets for each point.

This allows it to preserve important geometric information in the data that exists between points near each other on the manifold.

Returning to our code in Listing 5-7, let's add to our code and explore LLE mapping to a two-dimensional space with varying numbers of neighbors. For this package, you'll need to download the package (*https://mran.microsoft.com/snapshot/2016-08-05/web/packages/TDAmapper/README.html*) and locally install it:

```
#install package
library(devtools)
install_local("~/Downloads/lle.tar")

#create LLE projections of the data generated in Listing 5-6
library(lle)

l1<-lle(mydata,m=2,k=5)
l2<-lle(mydata,m=2,k=10)
l3<-lle(mydata,m=2,k=20)

#plot results
plot(l1$Y,xlab="Coordinate 1",ylab="Coordinate 2",main="K=5 LLE Results")
plot(l2$Y,xlab="Coordinate 1",ylab="Coordinate 2",main="K=10 LLE Results")
plot(l3$Y,xlab="Coordinate 1",ylab="Coordinate 2",main="K=20 LLE Results")
```

This piece of code applies the LLE algorithm to our dataset, varying the number of nearest neighbors considered in the algorithm calculations. Let's examine the plots from this dataset to understand the role of nearest neighbors in this local algorithm (Figure 5-24).

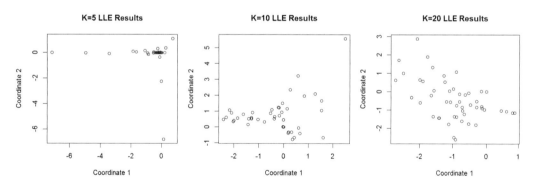

Figure 5-24: A side-by-side view of LLE results, which vary by number of nearest neighbors

As shown in Figure 5-24, neighborhood size greatly impacts LLE results and the spread of points in the new two-dimensional space. Given that the number of nearest neighbors impacts the size of the neighborhood preserved, higher values result in less-local versions of LLE, converting the algorithm into more of a global-type manifold learning algorithm. Good separation seems to occur at K=20, which is less local than K=5 but still a fairly small neighborhood for a dataset with 100 points. A fully global algorithm exists if we set K to 100, giving a two-dimensional plot with good separation and spread of points across the new space; you can see this in Figure 5-25.

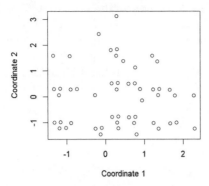

Figure 5-25: A plot of LLE results using the entire sample as nearest neighbors

Other local manifold learning algorithms exist, and some of these allow for a scaling parameters like LLE's neighborhood size. If you're interested, you can explore Laplacian eigenmaps, Hessian LLE, and local tangent space analysis.

Visualizing with t-Distributed Stochastic Neighbor Embedding

We've now seen how local algorithms can capture global properties through neighborhood size definition. Some manifold learning algorithms exist that explicitly capture both local and global properties. One of the more well-known algorithms is a visualization tool called *t-distributed stochastic neighbor embedding (t-SNE)*. The algorithm has two main stages: creating probability distributions over points in the high-dimensional space and then matching these distributions to ones in a lower-dimensional space by minimizing the Kullback–Leibler divergence over the two sets of distributions. Thus, rather than starting with a distance calculation between points, this algorithm focuses on matching distribution distances to find the optimal space.

Instead of defining a neighborhood by *k*-nearest neighbors to a point, t-SNE defines a neighborhood by the kernel's bandwidth over the data; this yields a parameter called *perplexity*, which can also be varied to understand the role of neighborhood size. Let's return to the data generated in Listing 5-7 and see how this works in practice. Add the following code:

```
#create t-SNE projections of the data generated in Listing 5-7
library(Rtsne)
library(dimRed)

t1<-getDimRedData(embed(mydata,"tSNE",ndim=2,perplexity=5))
t2<-getDimRedData(embed(mydata,"tSNE",ndim=2,perplexity=15))
t3<-getDimRedData(embed(mydata,"tSNE",ndim=2,perplexity=25))

#plot results
plot(as.data.frame(t1),xlab="Coordinate 1",ylab="Coordinate2",
```

```
main="Perplexity=5 t-SNE Results")
plot(as.data.frame(t2),xlab="Coordinate 1",ylab="Coordinate2",
main="Perplexity=15 t-SNE Results")
plot(as.data.frame(t3),xlab="Coordinate 1",ylab="Coordinate2",
main="Perplexity=25 t-SNE Results")
```

This piece of code runs t-SNE on the dataset generated with Listing 5-7, varying the perplexity parameter. The plots should produce something like Figure 5-26, which shows more clumping in the lower-perplexity trial than in the trials with higher perplexity.

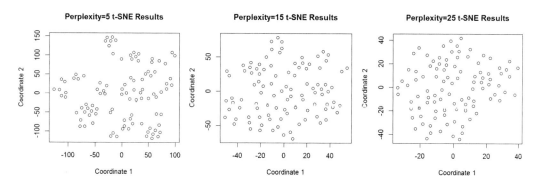

Figure 5-26: A side-by-side plot of t-SNE results with differing perplexity settings

The plots for perplexity of 15 and 25 look fairly similar, and as we increase perplexity, the range of the coordinates in the lower-dimensional space drops. There may be projects where more spread in the data is useful for subsequent analyses or visualizing possible trends; other projects may yield better results with tighter data.

In summary, the distance metrics in this chapter pop up regularly in machine learning applications. Manifold learning, in particular, can involve different choices of metric, neighborhood size, and type of space onto which the data space is mapped. Many good textbooks and papers exist that cover these algorithms and others like them in more detail. However, we hope that you've gained an overview of dimensionality reduction methods—particularly those that are intimately connected to metric geometry.

Before moving on, let's consider one final use of metric geometry.

Fractals

Another tool connected to metric geometry involves a type of self-similarity in geometry objects called *fractals*. Essentially, fractals have a pattern within the same pattern within the same pattern within the same pattern, and so on. Figure 5-27 has an example.

Figure 5-27: An example of a fractal. Note the self-similarity of the patterns at different scales.

Fractals occur often in natural and man-made systems. For instance, coastlines, blood vessels, music scales, epidemic spread in confined spaces, stock market behavior, and word frequency and ranking all have self-similarity properties at different scales. Being able to measure fractal dimension allows us to better understand the degree of self-similarity of these phenomena. There are many fractal dimension estimators these days, but most rely on measuring variations in the area under a fractal curve through some sort of iterative approach that compares neighboring point sets' areas.

Going back to the fractal in Figure 5-27, we could consider adding boxes to find the area under each iterative curve and then comparing the relative values across scales considered, as in Figure 5-28.

Figure 5-28: An example of measuring area under a series of fractal curves at scale

Now that we have some intuition around fractals, let's consider an application of fractal dimension metrics. Stock markets are known to exhibit some degree of self-similar behavior over periods of time. Understanding market volatility is a major aspect of investing wisely, and one method used to predict coming market reversal points, such as crashes, is changing self-similarity. The closing prices of the Dow Jones Industrial Average (one of the American stock market indices), or DJIA, are widely available for free download. Here, we'll consider simulated daily closing prices like DJIA data from the period of June 2019 to May 2020, during which the COVID freefall happened. Figure 5-29 shows a chart of closing prices over that time period.

Figure 5-29: A plot of simulated DJIA closing prices from June 2019 to May 2020. Note the big drops starting in late February 2020, when COVID became a global issue.

If we were predicting future market behavior, we'd want to employ fractal analyses with tools from time-series data analysis, which are outside the scope of this book. However, we can get a feel for changes in self-similarity month by month easily by parsing the data into monthly series and calculating each monthly series' fractal dimension. From there, we can examine how fractal dimension correlates with other measures of volatility, such as the range of closing prices within a month; we should see a positive correlation. Listing 5-8 loads the data, parses it, calculates fractal dimension, calculates closing price range, and runs a correlation test between fractal dimension and range.

```
#load and parse stock market data
stocks<-read.csv("Example_Stock_Data.csv")
June2019<-stocks[stocks$Month=="June",]
July2019<-stocks[stocks$Month=="July",]
August2019<-stocks[stocks$Month=="August",]
September2019<-stocks[stocks$Month=="September",]
October2019<-stocks[stocks$Month=="October",]
November2019<-stocks[stocks$Month=="November",]
December2019<-stocks[stocks$Month=="December",]
January2020<-stocks[stocks$Month=="January",]
February2020<-stocks[stocks$Month=="February",]
```

```
March2020<-stocks[stocks$Month=="March",]
April2020<-stocks[stocks$Month=="April",]
May2020<-stocks[stocks$Month=="May",]

#calculate fractal dimension for each series
library(fractaldim)
junedim<-fd.estimate(June2019[,2],methods="hallwood")$fd
julydim<-fd.estimate(July2019[,2],methods="hallwood")$fd
augustdim<-fd.estimate(August2019[,2],methods="hallwood")$fd
septemberdim<-fd.estimate(September2019[,2],methods="hallwood")$fd
octoberdim<-fd.estimate(October2019[,2],methods="hallwood")$fd
novemberdim<-fd.estimate(November2019[,2],methods="hallwood")$fd
decemberdim<-fd.estimate(December2019[,2],methods="hallwood")$fd
januarydim<-fd.estimate(January2020[,2],methods="hallwood")$fd
februarydim<-fd.estimate(February2020[,2],methods="hallwood")$fd
marchdim<-fd.estimate(March2020[,2],methods="hallwood")$fd
aprildim<-fd.estimate(April2020[,2],methods="hallwood")$fd
maydim<-fd.estimate(May2020[,2],methods="hallwood")$fd

#combine fractal dimension results into a vector
monthlyfd<-c(junedim,julydim,augustdim,septemberdim,octoberdim,novemberdim,
decemberdim,januarydim,februarydim,marchdim,aprildim,maydim)

#examine monthly stock price range
monthlymax<-c(max(June2019[,2]),max(July2019[,2]),max(August2019[,2]),
max(September2019[,2]),max(October2019[,2]),max(November2019[,2]),
max(December2019[,2]),max(January2020[,2]),max(February2020[,2]),
max(March2020[,2]),max(April2020[,2]),max(May2020[,2]))

monthlymin<-c(min(June2019[,2]),min(July2019[,2]),min(August2019[,2]),
min(September2019[,2]),min(October2019[,2]),min(November2019[,2]),
min(December2019[,2]),min(January2020[,2]),min(February2020[,2]),
min(March2020[,2]),min(April2020[,2]),min(May2020[,2]))

monthlyrange<-monthlymax-monthlymin

#examine relationship between monthly fractal dimension and monthly range
cor.test(monthlyfd,monthlyrange,"greater")
```

Listing 5-8: *A script that loads the simulated DJIA closing data, calculates fractal dimension and range of closing prices, and runs a correlation test to determine the relationship between fractal dimension and closing price range*

You should find a correlation of around 0.55, or a moderate relationship between closing price fractal dimension and closing price range, that is around the 0.05 significance level on the correlation test. Self-similarity does seem positively tied to one measure of market volatility. The fractal dimension varies by month, with some months' dimensionality being close to 1 and others' dimensionality being quite a bit higher. Impressively, the fractal dimension shoots up to 2 for March 2020.

Given that we only have 12 months' worth of data going into our test, it's worth noting that we still find evidence for a positive relationship between fractal dimension and range of closing prices. Interested readers with their own stock market data are encouraged to optimize the time frame windows and potential window overlap chosen to calculate the series of fractal dimensions on their own data, as well as investigate the correlations with other geometric tools used in stock market change point detection, such as Forman–Ricci curvature and persistent homology.

Summary

In this chapter, we've investigated metric geometry and its application in several important machine learning algorithms, including the k-NN algorithm and several manifold learning algorithms. We've witnessed how the choice of distance metric (and other algorithm parameters) can dramatically impact performance. We've also examined fractals and their relationship to stock market volatility. Measuring distances between points and distributions crops up in many areas of machine learning and impacts quality of machine learning results. Chapter 5 barely scratches the surface of extant tools from metric geometry. You may want to consult the papers referenced in the R packages used in this chapter, as well as current machine learning publications in distance metrics.

6

NEWER APPLICATIONS OF GEOMETRY IN MACHINE LEARNING

In Chapter 5, we explored the contributions of metric geometry to machine learning and its myriad uses in model measurements and input. However, geometry has provided many other contributions to machine learning; in this chapter, we'll explore tangent-space-based approaches to model estimation, exterior calculus, tools related to the intersection of curves (which can be used to replace linear algebra in algorithms), and rank-based models that involve vector fields acting on datasets' tangent spaces. We'll see how these tools can help in supervised learning on small datasets, help communities plan for disasters, and discern choice preferences of customers.

Working with Nonlinear Spaces

Our first tool helps mathematicians and machine learning engineers work with nonlinear spaces such as manifolds; it's the definition of a point's *tangent space*. Thinking back to calculus classes, we recall the *tangent lines* of a function are lines that touch a point on a curve without crossing the curve—where the slope of the curve equals the slope of the tangent line (giving the first derivative of the curve). Consider the sine wave example and a point on that sine wave, along with its tangent line, shown in Figure 6-1.

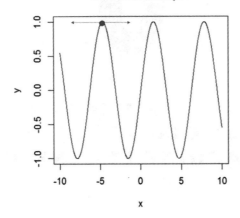

Figure 6-1: A sine wave example with tangent line drawn at one of the local maxima

This kind of tangent line works well in two dimensions. However, trying to define tangent lines to a point on a surface gets trickier, as many (infinitely many) lines can be tangent to a given point (see Figure 6-2, point A).

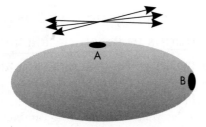

Figure 6-2: An ellipse with multiple possible tangent lines through point A

In fact, the lines in Figure 6-2 form a two-dimensional plane tangent to point A, akin to a sheet of paper that touches the ellipse at point A. What one could do is glue this *tangent plane* to point A, as in Figure 6-3.

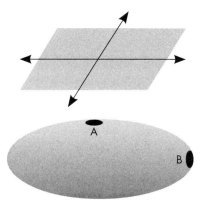

Figure 6-3: An ellipse with point A and the tangent plane associated with point A that extends tangent lines to tangent planes and spaces

In the case of higher-dimensional objects, the tangent spaces can grow to involve more dimensions (it can be 3-dimensional, 100-dimensional, or even an infinite-dimensional box). These tangent spaces have a nice connection to linear algebra. Remember that a vector space can be defined by a set of independent vectors collected into a matrix, called the *basis* of the space, technically known as *Hamel basis*. The basis for the tangent space of an object at a point, in fact, is the set of a point's partial derivatives. As mentioned earlier, in a one-dimensional space, this is exactly the slope of the tangent line. This gives a nice Euclidean space associated with each point on the manifold, which can be used to derive unit distances between points, provide mappings to a Euclidean space from a manifold, and understand multicollinearity. Multicollinearity occurs when variables are strongly correlated, which results in matrix columns or rows that are identical or nearly identical (causing singular matrices). Multicollinearity is a problem for regression-based algorithms, as it leads to redundant predictors and singular matrices. Variables with perfect overlap of variance (highly collinear predictors) will have the same tangent space or at least share some overlapping tangent space.

Introducing dgLARS

One useful machine learning algorithm based on tangent spaces is the *dgLARS* algorithm (dgLARS stands for "differential geometry least angle regression"). dgLARS extends traditional least angle regression (LARS) to an algorithm that fits to a given model's error tangent space. The LARS algorithm traditionally starts with each coefficient in the regression model set to 0, with predictors added progressively according to which predictor is most correlated with the outcome; coefficients are adjusted through least squares computation until a higher correlation enters the model. When multiple predictors have entered the model, coefficients are increased in joint least squares directions.

dgLARS considers the model's tangent space, scaling the score function used to optimize the coefficients. Each update to the model is done using the square root of a tool called the *conditional Fisher information*. The conditional Fisher information roughly measures the amount of information a given variable contains relative to a target (such as an outcome variable). For more technical-minded readers, the Fisher information of a parameter is the variance of the parameter's score, which is the partial derivative of that parameter with respect to the natural logarithm of the likelihood function.

Let's make this concrete with an example. Say we are creating a model to understand factors that impact adolescent risk-taking behaviors, such as drug use or petty crime. We may have many known factors measured already (such as family socioeconomic status, secondary school grades, and prior school or legal incidents). However, we'd like to compare an index from a survey we've designed to measure risk propensity in adolescents (index 1) to a known index that measures risk-taking in adults (index 2). Both surveys likely capture different types of information and different levels of relevant information with respect to our risk outcomes (Figure 6-4).

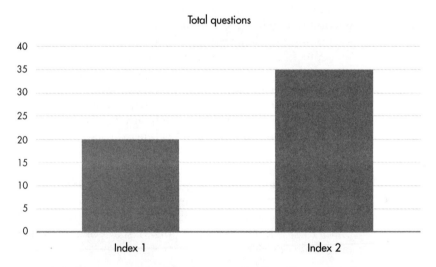

Figure 6-4: Comparison of number of questions loading onto a risk index

In the Figure 6-5 example, survey 1's index contains a greater volume of information. However, there may be some overlap between the variables or irrelevant information in survey 2's index, and it would be nice to have a measurement that can capture such information. Perhaps there is some overlap of information between the questions asked and some irrelevant information in each survey as well. Let's consider unique and relevant information contained in each risk-propensity survey (Figure 6-5).

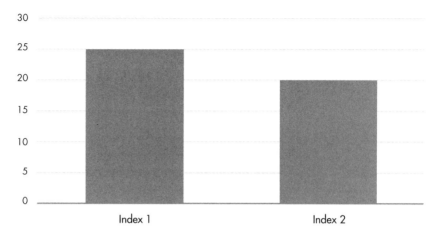

Figure 6-5: Two adjusted measures of relevant information captured in index questions

From Figure 6-5, we see that index 1 and index 2 contain some irrelevant information and some overlapping information. Index 1 does seem to contain a bit more information than index 2, and if we had to choose which survey to administer to a larger population of at-risk adolescents, we'd be better off starting with index 1.

This is a bit how Fisher information and variable selection in dgLARS happen. Technically, a *score* is calculated through partial derivatives of the model's log likelihood function, and the variance of this score is the Fisher information, which can be entered into a matrix to capture information across partial derivatives of the model. Interestingly, this matrix can also be derived as the Hessian of the relative entropy (Kullback–Leiber divergence), and the Fisher information gives the curvature of relative entropy in this case. At a less technical level, the Fisher information used to select variables has to do with both a statistical measure (the log likelihood function) and the information geometry of the independent variables being considered.

In the case of generalized linear models, the Fisher information matrix can be used to derive a score called the *conditional Rao score*, which can test whether the coefficient for a given variable is statistically different from 0. If the score is not statistically different from 0, the variable is dropped from consideration in the model. In the dgLARS algorithm, these calculations are done by searching coefficient vectors in the model error's tangent space, starting with the null model. This space's geometry is then iteratively partitioned into three sets: *selected predictors*, which have good fit scores in the error tangent space; *redundant predictors*, which share an error tangent space with selected predictors; and *nonselected predictors*, which have poor fit scores in the error tangent space. Thus, dgLARS leverages information about the model's geometry to find a best-fitting model.

dgLARS has had success on problems involving more predictors than observations in datasets, and many of the publications on this algorithm focus on genomics applications, where the number of patients might be 300 and the number of genes sequences might be in the 1,000,000 range. R provides a package, dglars, that implements this algorithm for generalized linear models, including link functions for logistic regression, Poisson regression, linear regression, and gamma regression. For those unfamiliar with generalized linear regression, a *link function* essentially is a special type of transformation of a dependent variable that allows regression to work mathematically for count variables, binary variables, and other types of not normally distributed dependent variables (with some restrictions on the geometry of the distribution).

Now that we've covered some of the theory, let's put it into practice.

Predicting Depression with dgLARS

We'd like to predict self-reported depression based on school issues and IQ. The data we'll use is self-reported school issues in a self-selected sample of profoundly gifted Quora users (with IQs above 155), including seven main school issues (bullying, teacher hostility, boredom, depression, lack of motivation, outside learning, put in remediation courses). The data was reported across 22 individuals who provided scores in the profoundly gifted range and discussed at least one of the issues of interest in a school system, with a bias toward users in the United States. This dataset was collected from the platform and processed manually to obtain a final dataset with categories of school issues for posters who met the IQ criterion. You can find the dataset in the book's files. Let's load the data first with the code in Listing 6-1.

```
#load data
mydata<-read.csv("QuoraSample.csv")
set.seed(1)
```

Listing 6-1: A script that loads the Quora dataset

The code in Listing 6-1 loads a dataset examining educational interventions and a psychological metric of self-esteem. Now let's modify the script to run the dgLARS algorithm on the dataset. The R package gives us the option of doing cross-validation or running the algorithm without cross-validation. Let's modify Listing 6-1 to run both model options:

```
#run analysis
library(dglars)
dg<-dglars(factor(Depression)~.,data=mydata,family="binomial")
dg1<-cvdglars(factor(Depression)~.,data=mydata,
family="binomial",control=list(nfold=2))
```

This runs the cross-validated and non-cross-validated dgLARS algorithms on the full set of student data. The cross-validated version does not work well on small datasets with sparse predictors, so if you run into an error, keep trying to run the cross-validated version, as some partitions may produce an error related to splitting and fitting models to the splits. The output from the two dgLARS algorithms should agree on many predictors, though the cross-validated split version of the algorithm may vary a bit, as the data is randomly partitioned. Let's add our script to look at the summary of the first model:

```
#examine results of non-cross-validated model
>summary(dg)
Summary of the Selected Model

    Formula: factor(Depression) ~ 1
     Family: 'binomial'
       Link: 'logit'

Coefficients:
    Estimate
Int. -1.5041
---

              g: 1.265
  Null deviance: 20.86
Residual deviance: 20.86
            AIC: 22.86

Algorithm 'pc' (method='dgLASSO')
```

No factors are selected as important predictors in this model. There are a few main reasons why no factors might be selected by the model, including the existence of subpopulations with opposite effects that "average out" the angles of the subpopulations, the existence of outliers that warp the geometry, or a true null effect for predictors. Model fit is reasonable, with an Akaike information criterion (AIC) of 20.86 and a residual deviance much smaller than the null deviance with no terms added to the model. The dispersion parameter tells us that the data fits reasonably well to the binomial distribution (see how g is close to 1) without problems in the distribution that sometimes occur in real-world data.

Now let's add to our script to look at the results from the cross-validated trials:

```
#examine cross-validated model
>summary(dg1)
Call: cvdglars(formula=factor(Depression) ~ ., family="binomial",
    data=mydata, control=list(nfold=2))

Coefficients:
                     Estimate
Int.                  -1.6768
Bullying               0.3066
Put.in.Remedial.Course 0.2292

dispersion parameter for binomial family taken to be 1

Details:
   number of non-zero estimates: 3
```

```
                cross-validation deviance: 12.33
                                        g: 0.9838
                                   n. fold: 2
```

Algorithm 'pc' (method='dgLASSO')

The cross-validated model should give a bit different result than the non-cross-validated model, suggesting the models are finding some consistency but not completely overlapping across folds. In the cross-validated model, profoundly gifted children put in remedial courses tend to have higher rates of depression, and children like the ones in this sample who are put in remedial courses and start showing signs of depression may benefit from this intervention. As you can see, bullying is also a potential issue leading to depression, suggesting something we likely already know: that bullying in general shouldn't be tolerated in a school for optimal mental health outcomes.

Given the small sample size, it's likely that a generalized linear regression model would struggle to estimate the coefficients. The necessary sample size for topology- and geometry-based linear models seems to be smaller than linear regression, and the consistent results on this problem suggest these models can work on very, very small data. However, there is still probably a minimum sample size needed for cross-validation, so it's best to avoid doing even these analyses if the sample size is less than 10.

Predicting Credit Default with dgLARS

To understand how dgLARS works on a larger dataset with a binary outcome (logistic regression) and more observations and variables, let's consider another dataset. The UCI credit default dataset includes 30,000 credit cases in Taiwan (late 2005) and 23 predictors of defaulting, including demographics (age, marriage status, education status, and gender), credit limit, and prior usage and payment information. The goals of our analysis are to figure out what predictors are related to whether an account ends up defaulting and to assess the model fit of our dgLARS model.

Let's get started on this with the code in Listing 6-2.

```
#load data
mydata<-read.csv("UCIDefaultData.csv")

#load library
library(dglars)

#run analysis, scaling the predictors such that big numbers don't
#result in large differences in coefficient values
#scale the predictor data
mydata1<-scale(mydata[,-c(1,25)])
mydata1<-cbind(mydata1,mydata$default.payment.next.month)
colnames(mydata1)[24]<-"default.payment.next.month"

#run the dglars function with and without cross-validation
dg<-dglars(factor(default.payment.next.month)~.,
```

```
data=as.data.frame(mydata1),family="binomial")
dg1<-cvdglars(factor(default.payment.next.month)~.,
data=as.data.frame(mydata1),family="binomial")
summary(dg)
summary(dg1)
```

Listing 6-2: A script that loads, processes, and analyzes the UCI credit default dataset with the dgLARS and cross-validated dgLARS algorithms

This should yield two models with coefficients for most predictors. The first model is the non-cross-validated model version (DG), and the second model is the cross-validated version (DG1). Table 6-1 shows results from our run of the code.

Table 6-1: Coefficients of Terms in the UCI Credit Default dgLARS Model

Column1	DG estimate	DG1 estimate
Int.	−1.47	1.45
LIMIT_BAL	−0.10	−0.09
SEX	−0.05	−0.04
EDUCATION	−0.08	−0.07
MARRIAGE	−0.08	−0.07
AGE	0.07	0.06
PAY_0	0.65	0.65
PAY_2	0.10	0.09
PAY_3	0.09	0.09
PAY_4	0.03	0.03
PAY_5	0.04	0.04
PAY_6	0.01	0.01
BILL_AMT1	−0.39	−0.15
BILL_AMT2	0.16	0.02
BILL_AMT3	0.09	0.02
BILL_AMT5	0.03	0.00
BILL_AMT6	0.02	0.00
PAY_AMT1	−0.22	−0.17
PAY_AMT2	−0.22	−0.19
PAY_AMT3	−0.05	−0.05
PAY_AMT4	−0.06	−0.06
PAY_AMT5	−0.05	−0.05
PAY_AMT6	−0.04	−0.03

Some of the biggest predictors of default include the prior month's billing and payment history. Those with lower usage (BILL_AMT1), lower payments (PAY_AMT1), and on-time payments (PAY_0) are less likely to default on payment in the following month. This makes a lot of sense, given that most lending metrics prioritize lending at the best rates to those with low loads of debt and a track record of on-time payment.

The cross-validated dgLARS model penalizes prior month usage and payment total less than the non-cross-validated model, suggesting that prior month on-time payment is more important than specific numbers. The AIC score on the first model is fairly large (27,924), but it is quite a bit smaller than the null model's AIC score (31,705), indicating a better fit than the null model even with several predictors included.

Now let's compare this model with logistic regression and compare the AIC fit statistics, adding the following to our script:

```
#run logistic regression
gl<-glm(factor(default.payment.next.month)~.,
data=as.data.frame(mydata1),family="binomial"(link="logit"))

#calculate AIC of the model
AIC(gl)
```

This snippet of code runs the logistic regression on the dataset and calculates the model's AIC. In this example, the AIC should come out to around 27,925, almost exactly that of the dgLARS models. This suggests a convergence of logistic regression and the dgLARS algorithm; at this large a sample size, this result is expected. Logistic regression is the typical tool for large sample sizes, and it doesn't seem that we get a gain from using dgLARS in this case. However, given the convergence on large sample sizes, it's likely that dgLARS gives quality results at the smaller sample sizes that won't work with logistic regression.

Applying Discrete Exterior Derivatives

Another useful tool that has come out of differential geometry is *discrete exterior derivatives*. Discrete exterior derivatives involve building up discrete shapes from lower-dimensional discrete shapes. In prior chapters, we examined the concept of homology, which counts the holes in a given object; technically, this is done by finding an object's boundaries at a specific dimension. For instance, consider the boundaries of a triangle (Figure 6-6).

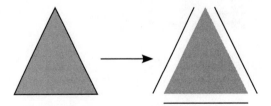

Figure 6-6: The boundaries of a triangle

We can take this a step further and break down lines into each point connected by the line, as in Figure 6-7.

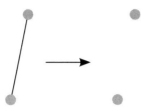

Figure 6-7: The boundaries of a line

Just as we can take apart shapes by identifying and separating out boundaries, we can also build shapes up from lower-dimensional boundaries by combining those boundaries. Technically, this is called *cohomology*, which is the realm of discrete exterior calculus. We might start with two points that are related in some way (perhaps within a certain distance of each other or sharing a characteristic) and connect them with a line (Figure 6-8).

Figure 6-8: Two points built into a line

For more technically minded readers, we're looking at the discrete version of differential forms, which are cochains on simplicial complexes. These differential forms have vector fields associated with them. We can then define operators that change those fields or objects, combine them, or count what exists within a field or cochain. This allows us to wrangle certain types of data to understand problems like resource capacity in electrical grids or burden of disease within social networks (or even rendering graphics across groups of pixels within a computer screen).

This can continue up to arbitrarily many dimensions, with lines building up triangles and triangles building up tetrahedra and so on. We can also jump levels with discrete exterior derivatives, going from points to triangles or tetrahedra rather than lines. Thus, discrete data, such as rendering pixel data or engineering data, can be grouped and connected for further study.

One of the newer applications of discrete exterior derivatives (and homology) is within social network analysis. As we mentioned in prior chapters, graphs are discrete objects of zero and one dimension (points and lines); however, connections between individuals can be extended from two-way mutual interactions (lines connecting points) to cliques of

3-way (triangle) or 4-way (tetrahedron) or 100-way mutual interactions (a very large dimensional sort of object), as demonstrated by three colleagues mutually connected in Figure 6-9.

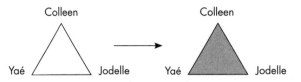

Figure 6-9: A graph of three colleagues working on projects together represented through a three-way interaction

On the left of Figure 6-9, we see three colleagues (Colleen, Jodelle, and Yaé) who collaborate in pairs but have not worked on a project involving all three of them. On the right of Figure 6-9, we see a representation of a project that involves all three colleagues working together. If they collaborate on many papers, we can sum up their two-way collaborations and their three-way collaborations to get a summary total for each n-way collaboration. This might be useful for understanding the strength of this collaborative network's parts.

Let's explore how these concepts can help in disaster logistics planning. Suppose there are four towns in a region with all towns connected to at least one other town by a road. Suppose also that each town has its own stock of supplies (perhaps liters of water for each resident) in case a cyclone hits the region and limits transportation for days or weeks. We can model this by creating a graph in R using the code in Listing 6-3.

```
#create matrix of town connections and miles between each town
towns<-matrix(c(0,0,0,4,0,0,12,2,0,12,0,6,4,2,6,0),nrow=4)

#create a graph from the matrix with connections between modes
#going in both directions (undirected graph) with weights
library(igraph)
g1<-graph_from_adjacency_matrix(towns,mode="undirected",weighted=T)

#plot town graph with edges labeled by weights
plot(g1,edge.label=E(g1)$weight,main="Plot of Connected Towns by Road Miles")

#add resource (perhaps liters of water per resident)
V(g1)$number<-c(10,500,80,200)
plot(g1,edge.label=E(g1)$weight,vertex.label=V(g1)$number,vertex.size=40,
main="Situation 1")

#find maximal cliques, representing connected resources
mc1<-max_cliques(g1)
mc1

#add resources that are mutually connected between towns
c1<-V(g1)$number[mc1[[1]][1]]+V(g1)$number[mc1[[1]][2]]
```

```
c2<-V(g1)$number[mc1[[2]][1]]+V(g1)$number[mc1[[2]][2]]+
V(g1)$number[mc1[[2]][3]]

#examine time needed to transport using shortest paths
dis1<-distances(g1,v=V(g1),mode=c("all"),weights=E(g1)$weight,
algorithm="dijkstra")
```

Listing 6-3: A script that generates the example graph of connected towns, plots the graph, adds resources to each town, visualizes these resources, and analyzes mutually connected town resources

Listing 6-3 creates a matrix of towns connected by roads and then converts this into a weighted graph. Once it is in graph form, we can add in information about resources available in each town and plot a picture with this information included, along with the distances between towns connected by a road. We can then calculate mutual resources between towns and minimum travel distances from a given town to another. This will help us assess resources available in a disaster and the best routes down which to send supplies.

Listing 6-3's first plot should output a diagram showing which towns are connected and the number of miles between towns, as shown in Figure 6-10.

Plot of Connected Towns by Road Miles

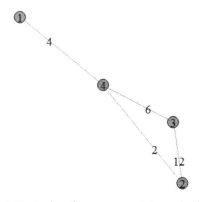

Figure 6-10: A plot of town connectivity and miles of road between connected towns

Figure 6-10 shows that towns 2, 3, and 4 are connected by multiple roads, such that if one road is blocked, the town can still be reached by looping around through another town. However, town 1 is relatively isolated despite being located only 4 miles from the nearest town. The road connecting towns 2 and 3 is quite long (perhaps this is a back road that meanders through a densely wooded area or around several canals).

The second plot in Listing 6-3 should output a graph that adds total resources for each town, as shown in Figure 6-11.

Situation 1

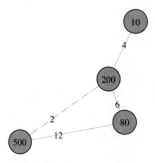

Figure 6-11: A plot of town connectivity, miles of road between connected towns, and resources within each town

Figure 6-11 gives us a richer understanding of connectivity and potentially shared resources among the four towns. Notice that town 1 has a relatively small water supply stocked for the disaster (10 liters per resident). However, if the road between towns 1 and 4 holds up during the disaster, it is easy to move some of the water from town 4 (with 200 liters per resident) to town 1, such that town 1 has sufficient water. It's also easy to move water between towns 4 and 2 (which has 500 liters per resident stocked up) provided the road directly connecting these towns holds up in the disaster.

The maximal clique calculation yields mutually connected towns (towns with mutual n-way connections). This gives a connection between towns 1 and 4 (with the single road) and towns 2, 3, and 4, which are mutually connected. From this information, we can calculate resources at each town's disposal should the roads hold between towns. Towns 1 and 4 mutually contain 210 resident-liters; the three-way clique (towns 2, 3, and 4) contains 780 resident-liters.

Using a shortest path algorithm, we can calculate shortest routes between any two towns to understand how quickly supplies might be routed between towns should supplies run low in a given town and roads are not damaged by the disaster. Table 6-2 gives the miles that would need to be driven between towns in this scenario.

Table 6-2: Shortest Distances Between Pairs of Towns

	Town 1	Town 2	Town 3	Town 4
Town 1	0	6	10	4
Town 2	6	0	8	2
Town 3	10	8	0	6
Town 4	4	2	6	0

The longest route in this scenario (10 miles) is trying to get supplies from town 3 to town 1 when town 1 runs low on water. Notice that it's shorter to bypass the longest road to route supplies between towns 2 and 3 (8 miles versus 12 miles).

We can examine a scenario where one or more roads is damaged in the disaster by adding to Listing 6-3 to remove roads connecting towns and re-examine mutual supplies and shortest paths:

```
#remove the link between vertices 2 and 4, providing an efficient supply
#route
g2<-delete_edges(g1,edges="2|4")
V(g2)$number<-c(10,500,80,200)
plot(g2,edge.label=E(g2)$weight,vertex.label=V(g2)$number,vertex.size=40,
main="Situation 2")

#find maximal cliques
mc2<-max_cliques(g2)
mc2

#find resources between sites
c1<-V(g2)$number[mc2[[1]][1]]+V(g2)$number[mc2[[1]][2]]
c2<-V(g2)$number[mc2[[2]][1]]+V(g2)$number[mc2[[2]][2]]
c3<-V(g2)$number[mc2[[3]][1]]+V(g2)$number[mc2[[3]][2]]

#examine time needed to transport using shortest paths
dis2<-distances(g2,v=V(g2),mode=c("all"),weights=E(g2)$weight,
algorithm="dijkstra")
```

This script removes a link between towns 2 and 4, recalculating the metrics to help us assess how the situation has changed with the blockage of the road between towns 2 and 4. These modifications should yield a plot of situation 2 similar to Figure 6-12.

Situation 2

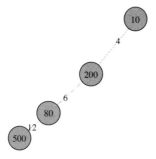

Figure 6-12: A modified plot of situation 1 with one road destroyed in the disaster

As Figure 6-12 shows, this scenario breaks the triangle present in situation 1, isolating town 2. If town 1's water supply runs low, the water must be shared between towns 1 and 2 with the expectation that it may be difficult to move water from town 2 to replenish the supplies.

If we look at the cliques, we can see two-way connections among towns, and mutual supplies between towns are more spread out than in the three-way connection present in situation 1. Towns 2 and 3 have a decent mutual supply, but the other towns may have delays in routing needed water. In addition, the miles needed to travel increases, as we can see in the shortest path table (Table 6-3).

Table 6-3: Shortest Distances Among Pairs of Towns

	Town 1	Town 2	Town 3	Town 4
Town 1	0	22	10	4
Town 2	22	0	12	18
Town 3	10	12	0	6
Town 4	4	18	6	0

Routing water from the town with the largest supply (town 2) has become a lot more difficult, with greatly increased expected travel times. Should scenario 2 appear likely, it's probably best to redistribute the water prior to the disaster to avoid delays in routing.

Discrete exterior derivatives have other applications. Graphics rendering and engineering problems have been particular areas of interest within discrete exterior derivatives (and discrete exterior calculus in general). A few of the more common applications in engineering are flux and flow calculations on discrete objects or computer modeling of processes. Within graphics rendering, graphs are typically replaced with more general meshes.

In some instances, it is easier to use cohomology (and its tool, discrete exterior derivatives) to study an object or point cloud than to use persistent homology, as the end result will find the same boundaries and objects. However, little has been done to make explicit R packages for applying discrete exterior derivatives to data, and code must be parsed together as in Listing 6-3. Automation and object manipulation code would help facilitate the adoption of discrete exterior derivatives within data science and other fields.

Nonlinear Algebra in Machine Learning Algorithms

Another intriguing and recent development with respect to geometry in machine learning is the notion of nonlinear algebra in machine learning algorithms. Many machine learning algorithms rely heavily on linear algebra to compute things such as gradients, least squares estimators, and so on. However, in relationships and spaces that aren't flat or involving straight lines, the linear algebra provides only an approximation of quantities calculated. Take a look at Figure 6-13, which shows a straight line (assumed by linear algebra tools, such as those used in regression) and a curved line (which may cause an estimation problem for linear algebra tools).

Figure 6-13: A straight line and curved line

Nonlinearity introduces error into the calculations and final result of an algorithm when curves and nonlinear spaces are estimated using linear tools. Imagine using a ruler to measure the lower line in Figure 6-13. It would be hard to get an exact length of the line relative to the straight line above. If the length of the line were a quantity that needed to be minimized or maximized by an algorithm, such a measurement could potentially find a nonglobal solution or distort the quantity enough to cause problems in predictive accuracy or model fit statistics like sum of square error or Bayesian information criterion.

One proposed alternative to linear algebraic calculations within machine algorithms is to use *numerical algebraic geometry*, a branch of nonlinear algebra that deals with the intersections of curves. For instance, consider the two-dimensional ellipse intersected by a one-dimensional curve, shown in Figure 6-14.

Figure 6-14: A plot of a curve intersecting an ellipse

Different types of matrices and matrix operations can be used to solve nonlinear systems analogously to how linear algebra solves linear systems; numbers populating a matrix are simply replaced with polynomial equations. Some of these intersecting curve problems are nonconvex problems, which often pose issues to machine learning algorithms and the linear algebra powering them.

Convex optimization problems are those in which the optimization function creates a region in which a line passing through the region is in the region continuously (rather than passing multiple regions and nonregions within the object), as shown in Figure 6-15.

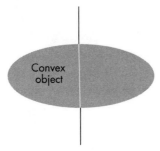

Figure 6-15: A convex object

If, however, the region contains holes or indents, this is no longer the case, and the region is designated as *nonconvex*, such as in Figure 6-16, where the hole inside the region splits the interior set into the set within the region and the set within the hole.

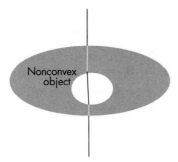

Figure 6-16: A nonconvex object

Optimization algorithms often struggle with nonconvex functions and regions within the optimization function, as the linear programming commonly used to solve these problems isn't as amenable to nonconvex optimization problems and as stepwise methods can get stuck in the local optima around the hole or divots in the region. Unfortunately, many real-world datasets and the optimization functions used on them involve nonconvex regions. Numerical algebraic geometry offers an accurate solution for nonconvex optimization problems, which come up often in real-world data situations. Once the system of polynomials is set up in matrix form, many software platforms and programs exist to do the computations, including the Bertini and Macauley software, which can connect to both R and Python. This allows for solvers that work well in nonconvex optimization problems.

Several recent publications and workshops in numerical algebraic geometry suggest that nonlinear algebra is a viable alternative to the current machinery in algorithms for other types of problems that may be convex, loosening mathematical assumptions and providing improvements in accuracy. One recent paper (Evans, 2018) found that the local geometry of many possible solutions overlaps for many types of statistical models (including Lasso, ARIMA models for time-series data, and Bayesian networks). This means that algorithms can't distinguish well between potential predictor sets or parameter values in the model space, particularly for the sample sizes commonly used and suggested for these problems. One can solve this problem by either fitting the model in tangent space, as we saw in the previous section, or using numerical algebraic geometry instead of linear algebra for optimization. This suggests some immediate applications of numerical algebraic geometry and other nonlinear approaches to machine learning algorithms for improved algorithm performance and model fits.

Unfortunately, packages to implement these algorithms do not exist in R at this time, so we won't explore this further in an example.

Comparing Choice Rankings with HodgeRank

Discerning choice preference across a population of customers is a common machine learning task. For instance, a company might want customers to compare items on a list of potential new features in software to prioritize the engineering team's time in the coming year. Comparing choice rankings also helps companies market new products and services to existing users and derive new campaigns for items that are likely to sell well with existing customers.

Let's look at a simple example of ranking activity preference for a day at the beach. Perhaps we're looking at data that a hotel collected from recent guests on which activities they preferred during their stay; this would allow them to prioritize beach usage and staff hiring to meet future demand for the main activities their guests prefer. Here, we have three choices of activity (lying on the beach, swimming, or surfing), with one preference as a clear favorite (surfing). Figure 6-17 summarizes this simple situation.

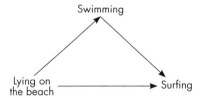

Figure 6-17: A diagram ranking three choices relative to the other choices

We can complicate this problem by adding a potentially new fourth activity that is preferred to the other three: kiteboarding. The ranking is still relatively easy to compute, as all activities are preferred to lying on the beach, one is preferable only to lying on the beach (swimming), one is preferred to every activity but kiteboarding (surfing), and one is preferred to the other three (kiteboarding), as shown in Figure 6-18.

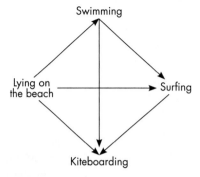

Figure 6-18: A diagram with another activity added to the preference data

All of the information is given in the diagram, with this particular person filling out all choices relative to each other. This is rarely the case with real-world data, as shown in Figure 6-19.

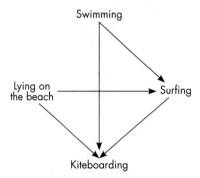

Figure 6-19: An incomplete preference diagram of the four activities

However, this information still shows a strong preference toward kiteboarding, with all pair-ranks existing for kiteboarding pairs and all pair-ranks pointing toward kiteboarding as the most preferred activity. In real-world data, it's common not only to have missing information but to have preference loops in the data, such as the one in Figure 6-20, where surfing is preferred to swimming, and kiteboarding is preferred to surfing but not to swimming.

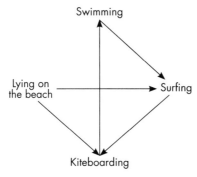

Figure 6-20: An example of incomplete and inconsistent ranking preference data

This gives a local inconsistency of where kiteboarding and swimming might rank relative to each other when other options are available. However, surfing and kiteboarding are preferred to two other activities, suggesting they rank toward the top of possible options, and kiteboarding is preferred to surfing (giving a tiebreaker of sorts).

The situation becomes a bit more nebulous when no consistent preferences exist globally, with each activity preferred to another activity, as shown in Figure 6-21.

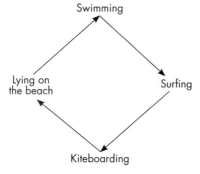

Figure 6-21: A diagram of incomplete preference data with no consistent preferences

In the case of Figure 6-21, nothing can be said about which activity would be preferred to other activities, and results of the analyses would be inconclusive. This is common when customers are asked to rank features in financial apps or guests are asked to rank preferred activities. Most customers will exist in subgroups with their own unique needs, which might be the opposite needs of another important customer subgroup. It's a challenge to prioritize features for development or choices to give guests to please the largest number of customers.

Many algorithms exist to do pairwise-rank comparisons to get a ranked list of preferences relative to each other (SVMRank, PageRank, and more); however, in general, they do not provide information about local and global inconsistencies in rank that might influence where an item or choice is placed relative to others. Algebraic geometry recently added a tool to the collection of pairwise-rank algorithms that can decompose the ranking results to include information about local and global inconsistencies of items; that would be *HodgeRank*, which can derive this information by leveraging an algebraic-geometry-based theorem common in engineering problems: the Hodge–Helmholtz decomposition.

The Hodge–Helmholtz decomposition partitions a vector flow (or flow on a graph) into three distinct components, shown in Figure 6-22: the *gradient flow*, which is locally and globally consistent; the *curl flow*, which is locally consistent but globally inconsistent; and the *harmonic flow*, which is locally inconsistent but globally consistent.

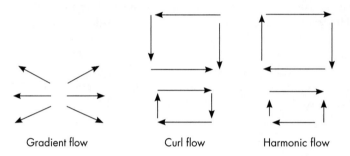

Gradient flow   Curl flow   Harmonic flow

Figure 6-22: A diagram showing the flows broken down by the Hodge–Helmholtz decomposition

In the beach examples, Figure 6-20 has a curl flow involving swimming, surfing, and kiteboarding (also harmonic if another activity is not ranked there). Figure 6-21 is an example of global harmonic flow (as well as local curl flow).

The HodgeRank algorithm essentially extends PageRank for pairwise-ranking problems; the math boils down to a least squares problem on the graph data, allowing for assessments of global ranking and local ranking consistency. Thus, suspicious rankings can be flagged for a human analyst's review. The algorithm also allows for a lot of missing data in the original pairwise ranking sets, making it widely applicable to the often-incomplete data on ranking problems in the real world (where users won't click through three million video options to rank each relative to all of the others!). While a package does not exist in R and thus we will not walk through an example, implementations do exist in Matlab, and readers familiar with Matlab who are interested in this algorithm are encouraged to use the resources listed for HodgeRank.

Summary

In this chapter, we learned about newer algorithms derived from differential and algebraic geometry and explored the use of both dgLARS and discrete exterior calculus on data analysis problems, including the Quora gifted sample, a credit-lending sample, and a disaster-planning scenario. Many more algorithms are being developed, and we've given an overview of how nonlinear algebra and Hodge theory have contributed to machine learning in recent years, impacting important types of industry problems (such as preference ranking and parametric model estimation).

In the next chapter, we'll return to persistent homology and examine another tool of algebraic topology called the Mapper algorithm. Both of these will be used on our student sample introduced in this chapter.

7

TOOLS FOR TOPOLOGICAL DATA ANALYSIS

In this chapter, we'll explore algorithms that have a direct basis in topology and use them to understand the dataset of self-reported educational data encountered in Chapter 6. The branch of machine learning that includes topology-based algorithms is called *topological data analysis (TDA)*. You already saw some TDA in Chapter 4, where we used persistent homology to explore network differences. Persistent homology has gained a lot of attention in the machine learning community lately and has been used in psychometric data validation, image comparison analyses, pooling steps of convolutional neural networks, and comparisons of small samples of data. In this chapter, we'll reexamine persistent homology and look at the Mapper algorithm (now commercialized by Ayasdi).

Finding Distinctive Groups with Unique Behavior

Previously, we used persistent homology to distinguish different types of graphs. Recall from Chapter 4 that persistent homology creates simplicial complexes from point cloud data, applies a series of thresholds to those simplicial complexes, and calculates a series of numbers related to topological features present within each thresholded simplicial complex. To compare objects, we can use Wasserstein distance to measure the differences in topological features across slices.

Persistent homology has many uses in industry today. *Subgroup mining*, where we look for distinctive groups with unique behavior in the data, is one prominent use. In particular, we're often searching for connected components with the zeroth homology groups, or groups that are connected to each other geometrically (such as clusters in hierarchical clustering). In psychometric survey validation, for example, subgroup mining allows us to find distinct groups within the survey, such as different subtypes of depression within a survey measuring depression.

Let's walk through a practical example of subgroup mining with persistent homology related to self-reported educational data from a social networking site. We'll simulate data and compare persistent homology results using the TDAstats package in R and single-linkage hierarchical clustering using the hclust() function in R (see Listing 7-1). We'll return to Chapter 6's example dataset of gifted Quora users self-reporting their school experiences (see the book files for this dataset). In this example, we'll split the sample into sets of 11 individuals so that we can compare the persistent homology results statistically to ensure our measurements don't vary across samples from our population of students. This provides a validation that our measurement is consistent across the population.

```
#load data and set seed
mydata<-read.csv("QuoraSample.csv")
set.seed(1)

#sample data to split into two datasets; remove the IQ scores
s<-sample(1:22,11)
set1<-mydata[s,-1]
set2<-mydata[-s,-1]
```

Listing 7-1: A script that loads the educational dataset and splits it into two sets to be explored with persistent homology

Now that we have our dataset, we can apply persistent homology to understand the clusters. Specifically, we're looking at the zeroth Betti numbers, which correspond to connected groups, and other topological features of the data—see Chapter 4 for a refresher.

First, we need to compute the Manhattan distances between each student in the social network; we'll use these to define the filtration.

Manhattan distances are often a go-to distance metric for discrete data. Add the following to your script:

```
#calculate Manhattan distance between pairs of scores
mm1<-dist(set1,"manhattan",diag=T,upper=T)
```

Next, we want to apply the persistent homology algorithm to the distance-based data to reveal the persistent features. Using the TDAstats package, we can then add code to compute the zeroth and first Betti numbers of this dataset, using a relatively low-filtration setting set as the largest scale for the approximation (this will give us larger clusters). Finally, we can plot the results in a persistence diagram and a plot of hierarchical clustering:

```
#create the Vietoris-Rips complex (turn data into a simplicial complex)
library(TDAstats)
d1<-calculate_homology(mm1,dim=1,format="cloud")
#plot persistence diagrams where circles=connected components, triangles=loops
plot_persist(d1)
#hierachical cluster
plot(hclust(mm1),main="Hierarchical Clustering Results, Distance Data")
```

The calculate_homology() function converts the point-cloud data from the distance dataset to a simplicial complex; we can then apply a filtration to identify topological features appearing and disappearing across the filtration. There are other methods that can create simplicial complexes from data, but the Rips complex in this package is one of the easiest to compute.

Using the previous code, we've plotted two figures. The call to plot _persist() should give something like Figure 7-1. You can see there that it appears one main group exists, along with possibly a subgroup or two at the lower filtration level; however, the subgroup may or may not be a significant feature, as it is near the diagonal.

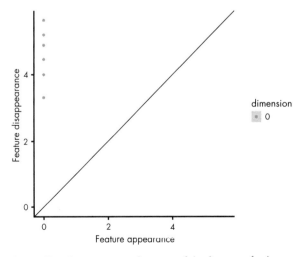

Figure 7-1: A persistence diagram of the first set of educational experience data

When using the hierarchical clustering results (Figure 7-2), it's easy to see a main group and then several splits at smaller distance thresholds.

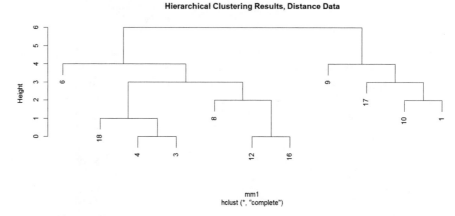

Figure 7-2: A dendrogram of the simulated data

If you cut the clusters at a height of 5, the dendrogram results suggest that two main subgroups exist. Let's split the set1 data according to the two main clusters found in the hierarchical clustering by adding to Listing 7-1, first examining the smaller cluster and then examining the larger cluster:

```
#smaller cluster
mydata[c(1,9,10,17),]
```

This should output the following:

```
> mydata[c(1,9,10,17),]
    IQ      Bullying Teacher.Hostility Boredom Depression Lack.of.Motivation
1   187     0               1             0         0              1
9   182     1               1             0         0              1
10  161     0               0             1         0              1
17  170     1               0             0         0              0
    Outside.Learning Put.in.Remedial.Course
1          0                    0
9          0                    1
10         0                    0
17         0                    0
```

In this cluster of individuals, no depression or outside learning was reported. Some individuals did report bullying, teacher hostility, boredom, remediation, or lack of motivation. Let's contrast that with the larger cluster found in our dendrogram:

```
#larger cluster
mydata[c(3,4,6,8,12,16,18),]
```

This should output something like this:

```
> mydata[c(3,4,6,8,12,16,18),]
   IQ Bullying Teacher.Hostility Boredom Depression Lack.of.Motivation
3  155        0                0       0          0                  0
4  155        0                0       0          0                  0
6  174        0                0       1          1                  1
8  170        0                0       0          0                  0
12 175        0                1       0          0                  0
16 160        0                1       0          0                  0
18 185        0                0       0          0                  1
   Outside.Learning Put.in.Remedial.Course
3                 1                      0
4                 1                      0
6                 1                      1
8                 0                      1
12                1                      1
16                1                      1
18                1                      0
```

Compared to the first cluster, these individuals mostly report outside learning and no bullying. This seems to separate learning experiences while in school. Learning outside of school and not dealing with bullying may have relevance to learning outcomes and overall school experience for students.

One item of interest in this analysis is individual 6, who seems to be an outlier in the Figure 7-2 dendrogram. This individual did not deal with bullying or teacher hostility but did deal with every other issue during their schooling. Outliers can be important and influential in analyses. Topology-based algorithms like persistent homology are often more robust to outliers than other algorithms and statistical models: extreme values or subgroups in the population won't impact the results as dramatically when we use TDA as compared to other methods. For instance, in our gifted sample, individual 6 might impact k-means results more than the results of persistent homology.

Subgroup mining is one important use of persistent homology—both for identifying groups within the dataset and for identifying outliers that might impact optimization steps in more traditional clustering methods. Let's continue by exploring another important use: as a measurement validation tool.

Validating Measurement Tools

Many methods exist to compare dendrograms or persistence diagrams; this is still an active area of research. Persistence diagrams need to be turned into metric spaces, which allows us to construct nonparametric tests with a compatible distance metric, which in turn lets us compare two diagrams and simulate random samples from the null distribution, which finally we can use to compare the test distance. All in all, this lets us validate measurement tools. In our example, we want to validate our measurement of school problems by comparing samples from the same population (our Quora sample). If persistent homology results are the same, our measurement tool is consistent, which is a key property of measurement design in the field of psychometrics.

For persistence diagrams, we typically use the Wasserstein distance, as it works well for comparing distributions and sets of points in finite samples. For dendrograms, Hausdorff and Gromov–Hausdorff distance are two good options, both of which measure the largest distance within a set of smallest distances between points on a shape. However, the Gromov–Hausdorff distance is more complicated and imposes more structural information, which makes it less ideal.

To compare the distances of another persistence diagram to the current one, let's use the second set of individuals in our self-reported educational dataset, adding to Listing 7-1:

```
#compute Manhattan distance for set 2
mm2<-dist(set2,"manhattan",diag=T,upper=T)

#create the Vietoris-Rips complex
d2<-calculate_homology(mm2,dim=1,format="cloud")

#plot persistence diagrams—circles=connected components, triangles=loops
plot_persist(d2)
```

Note that we've changed the dataset being analyzed to the second set of individuals from the full sample. This creates a comparison set that should be part of the same population; in this example, there are more potential subgroups that come out in the analysis. The plot should look something like Figure 7-3.

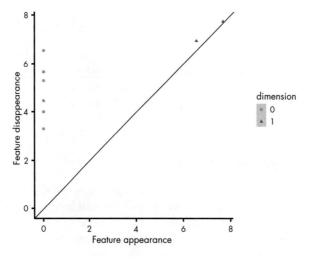

Figure 7-3: Another persistence diagram of simulated data, this time with different parameters used to simulate data

In Figure 7-3, we see a few distinct groups similar to Figure 7-1's sample. We also see some points corresponding to Betti number 1; however, given how close they are to the line, these points are likely noise. The farther from the diagonal line a point lies, the more likely it is a real feature in the dataset. These new Betti number features are different than our prior sample but likely not real features.

Computing the distance between the diagrams is easy with the TDAstats package. Add these lines to your script:

```
#calculate distance between diagrams
w<-phom.dist(d1,d2,limit.num=0)
w
```

This computes the distance between the persistence diagrams for the zeroth and first homology groups shown in Figure 7-1 and Figure 7-3 and should yield a distance of approximately 10.73 (zeroth homology) and 0.44 (first homology), though the values may vary according to your version of R. Now it's possible to compute the distances between random samples drawn from the original sample. The TDAstats package has a handy way of computing this within a function so that we don't have to write the entire test ourselves. Let's add these pieces to our script:

```
#compute permutation test and examine p-values from output
permutation_test(d1,d2,iterations=25)
```

This script will now compute a permutation test between the two samples' features, yielding a test statistic and p-value for each homology level computed. As expected, our zeroth homology differences are not significant at a 95 percent confidence level ($p = 0.08$). Given that we don't have any first homology features in our first sample, we do see a significant difference between the samples for our first homology differences; however, the statistic itself is 0, suggesting that this is an artificial finding.

While this example involves a convenience sample without an actual survey being administered, it does relate to how a real psychometric tool administered across population samples can be validated through persistent homology. We can also use this methodology to compare differences across different populations to explore how a measurement's behavior changes across populations. Perhaps one sample of students had been accelerated (skipped one or more grades) and one had not. We might end up with very different self-reported experiences in school. In this case, the measurement tool might show very different behavior across the proposed accelerated and nonaccelerated groups.

Using the Mapper Algorithm for Subgroup Mining

In data science, we often are faced with clustering problems where data has extreme variable scale differences, includes sparse or spread-out data, includes outliers, or has substantial group overlap. These scenarios can pose issues to common clustering algorithms like k-means (group overlap, in particular) or DBSCAN (sparse or spread-out data). The *Mapper algorithm*—which finds clusters through a multistage process based on binning, clustering, and pulling back the clusters into a graph or simplicial complex—is another useful clustering tool for subgroup mining. This algorithm ties together some of the concepts in Morse theory with the filtration concept in persistent homology to provide a topologically grounded clustering algorithm.

Stepping Through the Mapper Algorithm

The basic steps of the Mapper algorithm involve filtering a point cloud using a scalar-valued function called a *Morse function*; we then separate data into overlapping bins, cluster data within each bin, and connect the clusters into a graph or simplicial complex, based on overlap of the clusters across bins. To visualize this, let's consider a simple point cloud with a defined scalar-valued function; we'll shade the object according to the results we get when applying the function to the point cloud. Take a look at the results in Figure 7-4.

Figure 7-4: A multishaded object with a Morse function defined by a shade gradient

This shape can now be chunked into four overlapping bins. This allows us to see potentially interesting relationships between areas with slightly different Morse function values, which will become relevant when we apply a clustering algorithm, as in Figure 7-5.

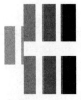

Figure 7-5: Binning results that chunk Figure 7-4 by shade gradient

Now that we've binned our function (Figure 7-5), we can start clustering. The clustering across bins can get a little bit more complicated than simply applying a clustering algorithm. This clustering is needed to define the complex and the overlapping of clusters across bins. Clustering within each of these bins and combining results to understand connectivity of clusters across bins would give a final result. An advantage of the Mapper algorithm is that results can be easily visualized as a graph or simplex; the final result of our example would likely output something like Figure 7-6, where two distinct groups evolve from a single point connecting them.

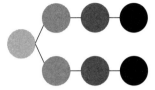

Figure 7-6: The clusters defined by binning the results of Figure 7-4

In practice, a distance metric—correlation, Euclidean distance, Hamming distance, and so on—is typically applied to the raw data before filtering as a way to process the point cloud data and create better filter functions prior to clustering. Clustering of the distance metric dataset can be done with a variety of algorithms, though single-linkage hierarchical clustering is usually used in practice. The coordinate systems used generally don't matter for Mapper results or results from other TDA algorithms.

There are a few advantages of the Mapper algorithm over other clustering methods, as well as topological data analysis compared to other methods in general. Invariance under small perturbations (noise) in the data allows Mapper to be more robust than k-means, which is sensitive to different starting seeds and can come up with very different results for each run. (Note that Mapper is sensitive to parameter changes but fairly robust to repeated runs with added noise.) The compression or visualization of results allows for easy visualization of clustering results for high-dimensional data. The lack of dependence on coordinate systems allows us to compare data on different scales or collected from different platforms. In addition, Mapper can deal with cluster overlap, which poses significant challenges to k-means algorithms and their derivatives. Lastly, Mapper's ability to handle sparsity and outliers gives it an advantage over DBSCAN. This makes it ideal for use on small datasets, datasets where predictors might outnumber observations, or messy data that is likely to contain a lot of noise.

Using TDAmapper to Find Cluster Structures in Data

The TDAmapper R package provides an implementation of the Mapper algorithm that can handle many types of processed data. For this example, we'll return again to the self-reported educational dataset from the sample of gifted Quora users, including seven main school issues (bullying, teacher hostility, boredom, depression, lack of motivation, outside learning, put in remediation courses) reported across 22 individuals who provided scores in the gifted range and discussed at least one of the issues of interest. The objective is to understand the relation between issues within this sample (somewhat like creating subscales within the measurement). This is binary data, so we'll use inverse Hamming distance to obtain a distance matrix. Hamming distance measures bit-by-bit differences in binary strings to get a dissimilarity measurement. Other distances can be used on binary data, but Hamming distance works well to compare overall differences between individuals scored on binary variables.

Let's load the data and prepare it for analysis in Listing 7-2:

```
#load data and clean
mydata<-read.csv("QuoraSample.csv")
mydata<-mydata[,-1]

#load Mapper, igraph, and distance packages
library(TDAmapper)
library(igraph)
library(e1071)

#process data to turn it into a distance matrix
df<-data.frame()
for (j in 1:7){
  for (i in 1:7){
    df[i,j]<-1/(hamming.distance(mydata[,i],mydata[,j]))
  }
}
df[df>1]<--1
```

Listing 7-2: A script that loads and processes the data to obtain a distance matrix

The code in Listing 7-2 loads our dataset and packages needed for the analysis and then processes the data to obtain a distance matrix to feed into the Mapper algorithm. Other distances can be used on binary data, but Hamming distance works well to compare overall differences between individuals scored on binary variables.

Now let's apply the Mapper algorithm. We'll set Mapper to process the distance matrix using three intervals with 70 percent overlap and three bins for clustering. A higher overlap parameter on a small dataset will encourage connectivity between clusters found across bins; in practice, a setting between 30 to 70 percent usually gives good results. In addition, the small number of intervals and bins correspond to about half the number of instances to be clustered in this dataset, which usually works well in practice. Generally, it's useful to use different parameter settings, as the results will vary depending on starting parameters; a few recent papers have suggested that the Mapper algorithm with nonvarying parameters is not wholly stable with respect to results. We'll also set Filter values according to minimum and maximum Hamming distances. We can do both by adding these lines to the script in Listing 7-2:

```
#apply mapper
j<-mapper1D(as.matrix(df),num_intervals=3,percent_overlap=70, num_bins_when_clustering=3,
filter_values=c(0.025,0.05,0.075,0.1,0.125,0.15,0.2))
summary(j)
j$points_in_vertex
```

This code runs the Mapper algorithm on the data with the parameters set earlier. The summary gives a list of objects in the Mapper object regarding results. The summary of points within a vertex gives us information as to how these variables separate into clusters.

Exploring the Mapper object yields some insight into which issues cluster together. We can gather a lot of information from the Mapper object, but this exploration will be limited to understanding which points from the dataset ended up in which cluster (vertex) in the Mapper object. Let's examine the output from our last addition to Listing 7-2:

```
$points_in_vertex
$points_in_vertex [[1]]
[1] 1

$points_in_vertex [[2]]
[1] 2

$points_in_vertex [[3]]
[1] 3

$points_in_vertex [[4]]
[1] 4

$points_in_vertex [[5]]
[1] 5

$points_in_vertex [[6]]
[1] 3 4 5 6

$points_in_vertex [[7]]
[1] 4 5 6 7
```

From the previous results, which show which variable shows up in which clusters, we can see that variables 1 and 2 (bullying and teacher hostility) tend to occur in isolation (points vertices 1 and 2), while other issues tend to occur in clusters (points in the remaining vertices). Given that these are authority-social and peer-social issues of social etiology rather than curriculum etiology, this makes some sense. How teachers interact and how students behave is typically independent of the curriculum, while issues stemming from lack of challenge in the classroom stem directly from a curriculum cause.

Adding to our script, we can plot in igraph to obtain a bit more insight into the connectivity of the clusters:

```
#create graph from the mapper object's adjacency information
g1<-graph.adjacency(j$adjacency,mode="undirected")
plot(g1)
```

This code turns the Mapper's overlapping cluster results into a graph object that can be plotted and analyzed to see how the clusters overlap with each other.

Figure 7-7 shows the isolation of the socially stemming issues of teacher hostility and bullying by peers. The curriculum-based issues tend to overlap to some extent with lack of motivation and outside learning (items 5 and 6) being the strong ties between these clusters.

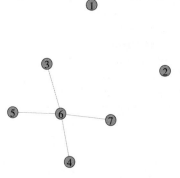

Figure 7-7: A network plot of the clusters found in the Quora sample analysis

One of the noted issues with Mapper is its instability with respect to overlap and binning of the filtration. For instance, changing the bin overlap to 20 percent results in the unconnected graph shown in Figure 7-8.

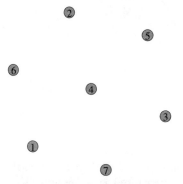

Figure 7-8: A network plot of the Quora sample results with a different parameter defining bin overlap

Some recent papers suggest using multiple scales to stabilize the output; however, most exploration of this notion is purely theoretical at this point. In general, using a variety of overlap fractions can yield a general idea of cluster structures in the data.

Summary

In this chapter, we explored a few tools from topological data analysis. We compared data from samples of an educational population using persistent homology and explored educational experience groups within a self-selected sample of profoundly gifted individuals. TDA has grown quite a bit in recent years, and many problems can be solved with one or more tools from TDA. In the next chapter, we'll explore one more popular tool from this growing field.

8

HOMOTOPY ALGORITHMS

In this chapter, we'll explore algorithms related to homotopy, a way to classify topological objects based on path types around the object, including homotopy-based calculations of regression parameters. Local minima and maxima often plague datasets: they provide suboptimal stopping points for algorithms that explore solution spaces locally. In the next few pages, we'll see how homotopy solves this problem.

Introducing Homotopy

Two paths or functions are *homotopic* to each other if they can be continuously deformed into each other within the space of interest. Imagine a golf course and a pair of golfers, one who is a better putter than the other. The ball can travel to the hole along many different paths. Imagine tracing out the path of each shot on the green with rope. One path might be rather direct. The other might meander quite a bit before finding the hole, particularly if the green is hilly. A bad golfer may have to make many shots, resulting in a long, jagged path. But no matter how many hills exist or how many strokes it takes for the golfer's ball to make it into the hole, we could shorten each of these paths by deforming the rope, as depicted in Figure 8-1.

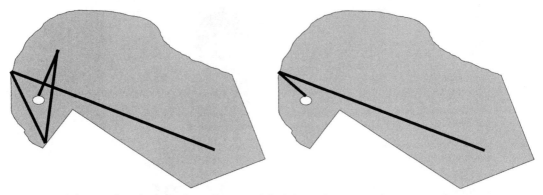

Figure 8-1: A long path to the hole of a golf course (left) deformed to a shorter path to the hole (right)

Let's stretch the analogy somewhat and imagine a sinkhole has appeared in the golf course. Topological objects and spaces with holes can complicate this deformation process and lead to many different possible paths from one point to another. Paths can connect two points on an object. Depending on the object's properties, these paths can sometimes "wiggle" enough to overlap with another path without having to cut the path into pieces to get around an obstacle (usually a hole). Winding paths around holes presents a problem to continuous deformation of one path into another. It's not possible for the path to wind or wiggle around a hole, such that a path between points will necessarily overlap with another path. Different types of paths begin to emerge as holes and paths around holes are added. One path might make only one loop around a hole before connecting two points. Another might make several loops around a hole before connecting two points. Imagine golfing again. Let's say that the green has an obstacle (such as a rock or a water hazard) in the middle of it, creating a torus with tricky hills around it that can force a bad shot to require circumnavigating the rock to get back to the hole, as you can see in Figure 8-2.

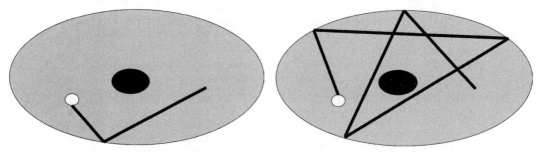

Figure 8-2: Two paths with the same start and end points on a torus course (donut) that cannot be morphed into each other without being cut or having the inner hole removed

In this scenario, we can no longer deform paths into each other without cutting the line or removing the hole. As more holes or holes of larger dimension are added, more classes of equivalent paths begin to emerge, with equivalent paths having the same number of loops around one or more holes. A two-dimensional course will have fewer possible paths from the tee to the hole than a three-dimensional course, as fewer possible types of obstacles and holes in the course exist. A space with many holes or obstacles in many different dimensions presents a lot of obstacles that paths can wind around. This means many unique paths exist for that space.

Given that datasets often contain holes of varying dimension, many different classes of paths may exist in the data. Random walks on the data, common in Bayesian analyses and robotic navigation path-finding tasks, may not be equivalent. This can be an advantage in navigation problems, allowing the system to choose from a set of different paths with different cost weights related to length, resource allocation, and ease of movement. For instance, in the path-finding problem in Figure 8-3, perhaps obstacle 2 has sharp ends that could harm the system should it get too close, making the leftmost path the ideal one for the system to take.

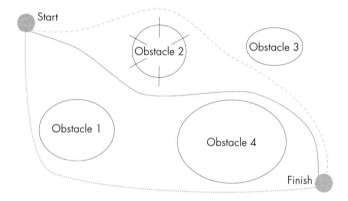

Figure 8-3: An example obstacle course with navigation from a start point to a finish point with several possible solutions

Figure 8-3 shows three paths, and none of them can be deformed into another of the paths without moving an obstacle or cutting the path. These are unique paths in the space. By counting the total number of unique paths, we can classify the space topologically.

Introducing Homotopy-Based Regression

As mentioned, datasets often contain obstacles in the form of local optima, that is, local maximums and minimums. Gradient descent algorithms and other stepwise optimization algorithms can get stuck there. You can think of this as the higher-dimensional version of hills and valleys (saddle points, which are higher-dimensional inflection points, can also pose optimization issues). Getting stuck in a local optimum provides less-than-ideal solutions to an optimization problem.

Homotopy-based algorithms can help with the estimation of parameters in high-dimensional data containing many local optima, under which conditions many common algorithms such as gradient descent can struggle. Finding a solution in a space with fewer local optima and then continuously deforming that solution to the original space can lead to better accuracy of estimates and variables selected in a model.

To provide more insight, consider a blindfolded person trying to navigate through an industrial complex (Figure 8-4). Without a tether, they are sure to bump into obstacles and potentially think they have hit their target when they are stopped by one of the larger obstacles.

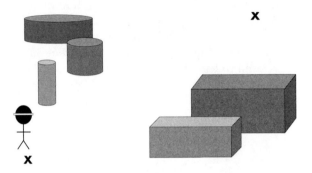

Figure 8-4: A blindfolded person navigating an obstacle course

However, if they are given a rope connecting their starting point to their ending point, they can navigate between the points a bit better and know that any obstacle they encounter is likely not the true ending point. There are many possible ways to connect the start and finish points. Figure 8-5 shows one possible rope configuration.

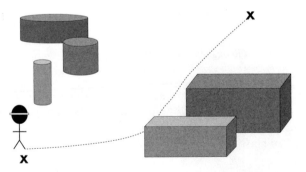

Figure 8-5: A blindfolded person navigating an obstacle course with a guide rope

A blindfolded person struggling to avoid physical obstacles is analogous to a machine learning algorithm avoiding local optima. For example, let's consider a function of two variables with a global maximum and minimum but other local optima, as derived in Listing 8-1.

```
#load plot library and create the function
library(scatterplot3d)
x<-seq(-10,10,0.01)
y<-seq(-10,10,0.01)
z<-2*sin(y)-cos(x)
which(z==min(z))
which(z==max(z))
scatterplot3d(x,y,z,main="Scatterplot of 3-Dimensional Data")
```

Listing 8-1: A script that creates a function of two variables with a global minimum and maximum but many other local optima

The code in Listing 8-1 produces the plot in Figure 8-6, from which we can see many minima and maxima. The other optima are local optima, some of which are very close to the global minimum or maximum.

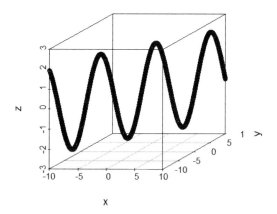

Figure 8-6: A scatterplot of three-dimensional data, namely, a function with many local optima

An algorithm trying to optimize this function will likely get stuck in one of the local optima, as the values near the local optima are increasing or decreasing from that optimum's value. Some algorithms that have been known to struggle with this type of optimization include gradient descent and the expectation-maximization (EM) algorithm, among others. Optimization strategies such as evolutionary algorithms will also likely take a long time to find global solutions, making them less ideal for this type of data.

Homotopy-based calculations provide an effective solution to this problem of local optima traps; algorithms employing homotopy-based calculations can wiggle around or out of local optima. In essence, these algorithms start with an easy optimization problem, in which no local optima are present, and deform the solution slowly according to the dataset and its geometry, avoiding local optima as the deformation proceeds.

Homotopy-based optimization methods commonly used these days in machine learning include support vector machines, Lasso, and even neural networks. The lasso2 package of R is one package that implements homotopy-based models; in this case, lasso2 implements a homotopy-based model for the Lasso algorithm. Let's first explore model fit and solutions for the data generated in Listing 8-1, in which the outcome has many local optima and the predictors are collinear, a problem for many machine learning algorithms. Add the following to the code in Listing 8-1:

```
#partition into training and test samples
mydata<-as.data.frame(cbind(x,y,z))
set.seed(10)
s<-sample(1:2001,0.7*2001)
train<-mydata[s,]
test<-mydata[-s,]
```

Now, the model is ready to be built and tested. The outcome of interest (our variable z) is not normally distributed, but a Gaussian distribution is the closest available distribution for use in the model. In the following addition to the script in Listing 8-1, the etastart parameter needs to be set to null before starting the model iterations, and a bound needs to be in place to guide the homotopy-based parameter search. Generally, a lower setting is best:

```
#load package and create model
library(lasso2)
etastart<-NULL
las1<-gl1ce(z~.,train,family=gaussian(),bound=0.5,standardize=F)
lpred1<-predict(las1,test)
sum((lpred1-test$z)^2)/601
```

This script now fits a homotopy-based Lasso model to the training data and then predicts test data outcomes based on this model, allowing us to assess the model fit. The mean square error for this sample, calculated in the final line, should be near 2.30. (Again, results may vary with R versions, as the seeding and sampling algorithms changed.) The results of the model suggest that one term dominates the behavior of the function:

```
> las1
Call:
gl1ce(formula=z ~ .,data=train,family=gaussian(),standardize=F,
    bound=0.5)

Coefficients:
(Intercept)              x               y
 0.05068903      0.04355682      0.00000000
```

```
Family:

Family: gaussian
Link function: identity


The absolute L1 bound was      : 0.5
The Lagrangian for the bound is : 1.305622e-13
```

These results, which may vary for readers with different versions of R, show that only one variable is selected as important to the model. x contributes more to the prediction of z than y contributes, according to our model. Linear regression isn't a great tool to use on this problem, given the nonlinear relationships between x or y and z, but it does find some consistency in the relationship.

To compare with another method, let's create a linear regression model and add it to Listing 8-1:

```
#create linear model
l<-lm(z~.,train)
summary(l)
lpred<-predict(l,test)
sum((lpred1-test$z)^2)/601
```

This code trains a linear model on the training data and predicts test set outcomes, similar to how the homotopy-based model was fit with the previous code. You may get a warning with your regression model, as there is covarying behavior of x and y (which presents issues to linear regression models per the assumption of noncollinearity). Let's take a look at this model's results:

```
> summary(l)
Call:
lm(formula=z ~ .,data=train)

Residuals:
    Min      1Q  Median      3Q     Max
-2.51261 -1.48663  0.07368  1.48680  2.37086

Coefficients: (1 not defined because of singularities)
            Estimate Std. Error t value Pr(>|t|)
(Intercept) 0.050689   0.041267   1.228    0.22
x           0.043557   0.007112   6.124 1.18e-09 ***
y                 NA                  NA
---
Signif. codes:  0 '***' 0.001 '**' 0.01 '*' 0.05 '.' 0.1 ' ' 1

Residual standard error: 1.544 on 1398 degrees of freedom
Multiple R-squared:  0.02613,   Adjusted R-squared:  0.02543
F-statistic: 37.51 on 1 and 1398 DF,  p-value: 1.183e-09
```

The mean square error (MSE) for this sample should be near 2.30, which is the same as the homotopy-based model. MSE accounts for both variance and bias in the estimator, giving a balanced view of how well the algorithm is performing on a regression task. However, the collinearity is problematic for the linear regression model. Penalized models avoid this issue, including homotopy-based Lasso models. Of note, the coefficients found by the linear regression and the homotopy-based Lasso model are identical. Typically, models with different optimization strategies will vary a bit on their estimates. In this case, the sample size is probably large and the number of predictors few enough for both algorithms to converge to a global optimum.

Comparing Results on a Sample Dataset

Let's return to our self-reported educational dataset and explore the relationships between school experiences, IQ, and self-reported depression. Because we don't know what the function between these predictors and depression should be, we don't know what sort of local optima might exist. However, we do know that a training dataset with 7 predictors and 16 individuals (70 percent of the data) will be sparse, and it's possible that local optima are a problem in the dataset. There is evidence that geometry-based linear regression models work better on sparse datasets than other algorithms, and it's possible that our homotopy-based Lasso model will work better, as well.

Let's create Listing 8-2 and partition the data into training and test sets.

```
#load data and set seed
mydata<-read.csv("QuoraSample.csv")
set.seed(1)

#sample data to split into two datasets with stratified sampling
#to ensure more balance in the training set with respect to depression
m1<-mydata[mydata$Depression==1,]
m2<-mydata[mydata$Depression==0,]
s1<-sample(1:4,3)
s2<-sample(1:18,6)
train<-rbind(m1[s1,],m2[s2,])
test<-rbind(m1[-s1,],m2[-s2,])
```

Listing 8-2: A script that loads and then analyzes the Quora IQ sample

Now, let's run a homotopy-based Lasso model and a logistic regression model to compare results on this small, real-world dataset:

```
#run the homotopy-based Lasso model
las1<-gl1ce(factor(Depression)~.,train,family=binomial(),bound=2,standardize=F)
lpred1<-round(predict(las1,test,type="response"))
length(which(lpred1==test$Depression))/length(test$Depression)
```

```
#run the logistic regression model
gl<-glm(factor(Depression)~.,train,family=binomial(link="logit"))
glpred<-round(predict(gl,test,type="response"))
length(which(glpred==test$Depression))/length(test$Depression)
```

From running the models in the previous script addition, we should see that the homotopy-based Lasso model has a higher accuracy (~85 percent) than the logistic regression (~70 percent); additionally, the logistic regression model spits out a warning message about fitted probabilities of 0 or 1 occurring. This means the data is quite separated into groups, which can happen when small data with strong relationships to an outcome is split. Depending on your version of R or GUI, you may end up with a different sample and, thus, somewhat different fit statistics and results. Because this is a relatively small sample to begin with, it's possible that you'll have slightly different results than the ones presented here. Some samples may not have any instances of a given predictor within the dataset. Larger samples would create more stable models across samples.

Let's look more closely at the homotopy-based Lasso model and its coefficients:

```
> las1
Call:
glice(formula=factor(Depression) ~ .,data=train,family=binomial(),
    standardize=F,bound=2)

Coefficients:
         (Intercept)                        IQ                Bullying
          -8.31322182                0.04551602              0.00000000
     Teacher.Hostility                   Boredom      Lack.of.Motivation
           0.00000000                0.51213722              0.00000000
     Outside.Learning  Put.in.Remedial.Course
          -1.01281345                0.42953330

Family:

Family: binomial
Link function: logit


The absolute L1 bound was       : 2
The Lagrangian for the bound is : 0.4815216
```

From the previous output, we can see that, for this sample, higher IQ, endorsement of boredom, and being put in a remedial class increase the likelihood of self-reported depression. However, outside learning has a strong protective effect. In fact, outside learning can completely counterbalance the risk from boredom and being placed in a remedial course. This suggests that parents of profoundly gifted children who are experiencing school issues may be able to mitigate some of the potential adverse outcomes, such as depression, by providing outside learning opportunities, such as college courses in the evening, tutoring outside of school, or other opportunities for the child to learn. The role of outside learning opportunities has been explored to some extent in the giftedness literature with similar results, but more research is needed on this topic.

Now, let's compare these results with the results of the logistic regression model:

```
> summary(gl)
CALL:
glm(formula=factor(Depression) ~ .,family=binomial(link="logit"),
    data=train)

Deviance Residuals:
        20          21           6          16          18           4
 3.971e-06   1.060e-05  -3.971e-06  -3.971e-06  -2.110e-08
        11           2          19
-8.521e-06  -1.060e-05  -1.060e-05

Coefficients:
                      Estimate  Std. Error  z value  Pr(>|z|)
(Intercept)          -7.297e+02   1.229e+06   -0.001         1
IQ                    3.934e+00   6.742e+03    0.001         1
Bullying             -1.421e+01   5.193e+05    0.000         1
Teacher.Hostility     2.798e+01   3.424e+05    0.000         1
Boredom              -1.967e+01   8.765e+04    0.000         1
Lack.of.Motivation    4.174e+01   2.496e+05    0.000         1
Outside.Learning     -6.535e+01   2.765e+05    0.000         1
Put.in.Remedial.Course 1.121e+02  2.712e+05    0.000         1

(Dispersion parameter for binomial family taken to be 1)

    Null deviance: 1.1457e+01  on 8  degrees of freedom
Residual deviance: 5.6954e-10  on 1  degrees of freedom
AIC: 16
Number of Fisher Scoring iterations: 24
```

Examining the previous output, it seems that the logistic regression model could not handle the dataset, giving errors and spitting out very large coefficients. This is likely related to the smallness of the data, where the linear system is underdetermined; however, this is not a situation where the number of predictors outnumber the number of individuals in the sample, so it is likely a function of the data itself rather than purely sample size.

Note the model fails to find any significant predictors of self-reported depression. Linear regression can't handle this dataset very well, and the results are not reliable. For some samples, certain variables may not be computable in the linear regression model at all. Homotopy-based models (and other types of penalized models) often work better on small datasets, and there is some evidence that they perform better for datasets with many local optima. While this dataset is a bit small for fitting a model, it does demonstrate the power of homotopy-based optimization (and penalized regression, in general) on very small datasets, and its results make a lot more sense than the linear regression model's results.

Summary

In this chapter, we gave you an overview of homotopy and its applications in machine learning, including through an example of homotopy as an extension of regression-based algorithms on a simulated problem and a real dataset. Homotopy can help regression algorithms avoid local optima that often trap local optimizers.

Other uses of homotopy algorithms in the field of artificial intelligence include navigational problems. For instance, an autonomous cart may need to navigate the halls and obstacles of a hospital by weighting different possible paths from its current location to its destination. Homotopy algorithms are often used to generate the possible paths, which are then weighted by time cost or hazard cost of the route. Bounds can also be placed to avoid generating paths that obviously aren't viable (such as going through areas where the cart can't physically go or wouldn't be wanted—such as an operating room). It's likely that this branch of topological data analysis will grow in the coming years, and we encourage you to explore other uses of homotopy in machine learning, robotics, differential equations, and engineering.

9

FINAL PROJECT: ANALYZING TEXT DATA

In this chapter, we'll put together some of the tools developed in previous chapters by working on a project related to linguistics and psychology. Many important data projects today deal with text data, from text matching to chatbots to customer sentiment analysis to authorship discernment and linguistic analysis. We'll look at a small dataset of linguistic features derived from different creative writing samples to see how language usage varies over genres; some genres encourage a different writing process that reflects a different author mindset, such as writing a personal essay versus writing a spontaneous haiku.

Specifically, we'll look at cluster overlap using k-NN and our distance metric of choice, then visualize the feature space by reducing the dimensionality of the dataset, then use dgLARS to create a cross-validated model distinguishing poetry types by features, and finally examine a predictive model based on language embedding. By completing a whole project, we'll see how these tools can fit together to derive insight from data.

Building a Natural Language Processing Pipeline

In Chapter 1, we briefly discussed the importance of text data and how natural language processing (NLP) pipelines can transform text data into model features that fit well with supervised learning methods. This is the approach we'll take with our upcoming project. We'll use Python to transform the text data into features and then use R to analyze those features.

The first step is to *parse* the data; we have to break the blocks of text into more manageable chunks—either sentences within a paragraph or words and punctuation within a sentence. This allows us to analyze small pieces of text and combine the results into a sentence, a document, or even a set of documents. For instance, in this chapter, you might want to parse each section, then each paragraph, and then each word in each paragraph to understand how the language usage varies between introductory material and application examples.

Sometimes, you'll want to take out punctuation, certain types of filler words, or additions to root words. *Root words* exclude endings like *ing*, which change the tense of a verb or turn one part of speech into another. Consider the differences between *dribble*, *dribbled*, and *dribbling* with respect to a basketball practice. If the point guard is dribbling the ball, the action is occurring now. If they've already dribbled for an entire game, they are probably tired and have put the ball back on the rack. However, the point guard's action, whether past or present, is the same. It doesn't matter much if they're doing it now or did it yesterday. Stemming and lemmatizing are two approaches that break words into root words: *stemming* does this by reducing words to their roots regardless of whether the root is still a word; *lemmatizing* reduces words to roots in a way that ensures the root is still a word.

Once text is parsed to the extent necessary for your specific application, you can start analyzing each piece. *Sentiment analysis* aims to understand the emotions behind the words and phrases of a given piece of text. For instance, "terrible product!!! never buy" has a fairly negative tone compared to "some users may not like how the product smells." Sentiment analysis quantifies emotions within the text by tallying up totals for each emotion within the text, such that each receives a score and can be rolled into a final score, if preferred.

Once parsing is done, we can apply *named entity recognition*, which matches words to lists of important people, places, or terms of interest in a field. For instance, when processing medical notes related to patient discharge and outcome, you might want to match words to a list of diagnoses.

In other applications, it may be important to tag parts of speech, including pronouns, verbs, prepositions, adjectives, and more. Some people might use certain parts of speech at higher rates than others, and understanding these patterns can give insight into the text source's personality, temporary state of mind, or even truthfulness. For some NLP applications, it's possible to load these factors onto a given outcome or set of outcomes to create entirely new metrics from text data.

Each of these analysis types can be integrated into a relational database as additional features for downstream models. For instance, as we'll see later in the chapter, we can tag parts of speech, normalize them by the length of that particular document, and feed those features into models. You can also vectorize the words that exist in the document or set of documents to count frequencies of each word that exists in the set of documents for a given document. This often precedes deep learning models and visualization methods within NLP applications.

Again, because R has limited NLP capabilities compared to Python, we've done this first step—parsing the text data into features—in Python and provided the resulting data in the files for the book. We'll then do the analysis using R. We used Python's NLTK toolkit to parse the text data, and while these steps are beyond the scope of this book, we have included the scripts in our Python downloadable repository (*https://nostarch.com/download/ShapeofData_PythonCode.zip*), and we encourage you to take the raw data provided and see if you can build a similar NLP pipeline.

The Project: Analyzing Language in Poetry

Modern poetry includes many types of poems with different structures, literary devices, and subject matters. For instance, formal-verse poems, such as sonnets or villanelles, have a defined number of stressed and total syllables in each line and a defined rhyme scheme. These types of poems often make heavy use of other literary devices, such as allusion or conceit (reference to other works as a juxtaposition of ideas), alliteration or assonance (repetition of a certain sound), or meter (patterns of stressed syllables within a line of poetry). This is an example of formal verse (a sonnet):

Ever After

Her glass slipper turns to his M-16,
her elegant dress to faded fatigues.
He's a shell of the man from their intrigues,
served five months patrol away from his queen.

Cinders to palace, her dreams now rubble,
she watches her carriage morph to Abrams tank,
as if her fairy tale were some cruel prank.
He shouts, "Hurry, men! March on the double!"

Her clock strikes twelve, his tank an IED,
and widower's daughter is left widow.
She has but memories, now as shadows,
to comfort her dark days of misery.
Her ever after has no tomorrow,
leaving Cinderella grief and sorrow.

In contrast, free-verse poetry doesn't have a defined rhyme scheme for the end of each line, may have varying line lengths (or consistently long, short, or medium lines), and tends to use literary devices such as meter or rhyme for emphasis of a particular piece of the poem. This is an example of a free-verse poem with short line lengths:

> Anya
>
> Gaunt,
> made-up,
> wobbling in heels
> too big for tiny feet,
> heels
> that sparkle and clack
> against cold concrete
>
> as wind
> whips her teased hair
> like a lasso
> roping a steer
>
> as snow
> binds to tight jeans
> like the shackles
> she wore on the ship to her Shangri La
>
> as streetlight
> catches a gleam in her eyes,
> a glimpse as she stares into
> tonight
>
> her fifteenth birthday.

Some poems don't fit neatly into free verse or formal verse, such as prose poetry (where line breaks don't exist) or haibun (a Japanese form that incorporates prose poetry with different types of haiku). Modern haiku and its related forms juxtapose two images or thoughts with a turn, such as a dash, that connects the two images or thoughts in a moment of insight. Typically, modern haiku doesn't conform to the Japanese syllable count requirements, but it usually includes some reference to season and nature (or human nature in the case of the related form, senryu). Haibun knits a story together through the poem title, the haiku, and the prose pieces. Here's one example of a modern haiku:

> shooting stars—
> the dash between
> born and died

From prior research, we know that different authors can be identified by their preferred word choices and their unique usage of different parts of speech (this is a core feature of antiplagiarism software); we also know that language usage varies by the author's state of mind. However, it's unknown if the same author constructs poetry differently depending on the type of poem they are writing. Because haiku and haibun originated as a spark of insight juxtaposing ideas, it's possible that the different genesis of the poem influences the use of language and grammar within the poem.

Let's dive into the dataset a bit before we start visualizing it. We have eight haiku, eight haibun, eight free-verse poems, and eight formal-verse poems in the dataset. We'll group them into haiku-based poems and other poems to simplify the analyses by how poems are typically generated (free association versus crafting). The features we'll consider are punctuation fraction, noun fraction, verb fraction, personal pronoun fraction, adjective fraction, adverb fraction, and preposition/conjunction fraction. Given the construction of each poem type, it's likely some of these factors will vary. With a larger sample of poems, you could use other parts of speech or break categories, such as verbs, into their individual components.

We'll be completing these steps in Python, so we'll overview only the steps used rather than dive into code. You can find the processed data in the files for this book.

Tokenizing Text Data

The first step in processing the poems for analysis is *tokenizing* the text data, meaning we need to parse our poems into individual words and punctuation marks. The Treebank tokenizer uses regular expressions to parse words in sentences, handle contractions and other combinations of words and punctuation, and splice quotes. These are important steps for parsing poem text data, as punctuation is interspersed with words and phrases at relatively high rates. Haiku, in particular, tends to use punctuation to create a cut in the poem to link two different images or ideas.

Because the Treebank tokenizer often splits contractions and other words that connect with punctuation, it's useful to the regex tokenizer to count the number of actual words that exist in the text and parse them into words that can contain punctuation. Given how short some poems are, we want to make sure we aren't inflating word counts. The regex tokenizer results give us an accurate word count to normalize parts of speech or punctuation proportions.

After obtaining the lengths of tokenizer results, we can subtract the number of words from the number of tokens to derive a length of punctuation in the text. This allows us to compare fractions of words and fractions of punctuation for different poem types, which likely varies by poem type (and by poem author, according to prior research on linguistic differences in text passages by author).

Tagging Parts of Speech

To tag relevant parts of speech, we'll use the *averaged perceptron tagger*, a supervised learning algorithm that tends to have pretty good accuracy across text types and has been pretrained for the NLTK package. While it's a bit slow on large volumes of text, our text samples are fairly small, allowing the application to tag words without much processing power required.

It's possible to scale NLTK's averaged perceptron tagger application to very large datasets using the big data technologies that we'll consider in the next chapter.

We'll parse out nouns, verbs, personal pronouns, adverbs, adjectives, and prepositions and conjunctions and count the numbers of each category that exist in each text sample.

Nouns include singular, plural, common, and proper noun combinations. Verbs include all type and tense combinations. Personal pronouns include pronouns and possessive pronouns. Adverbs and adjectives include comparative and superlative forms of adverbs and adjectives. Prepositions and coordinating conjunctions are also tagged and counted.

Some other tagged parts of speech exist in the averaged perceptron tagger, and other taggers may include further divisions of parts of speech. If you want to explore how parts of speech can be used to distinguish text types, text authorship, or demographics of the text author, you may want to use another tagger or disaggregate nouns, verbs, and so on, from their individual components. However, this will result in more columns within your dataset, so we recommend you collect more samples if you're doing that type of nuanced analysis of text attributes and parts-of-speech analysis.

Normalizing Vectors

Because our text samples include some short samples and some long samples according to poem type, we'll want to standardize the parts-of-speech counts before we work with the data. Our approach includes normalization of punctuation by token count (punctuation and word count totals) and normalization of parts-of-speech count by word count, summarized in this chapter's files. This should give good enough features to demonstrate poem type differences and still allow for good dimensionality reduction results to visualize our dataset.

For a more nuanced approach that parses out types of verbs, nouns, and so on, you could derive fractions of part of speech category or break down fractions within parts of speech to engineer more detailed features for your analysis. If you're familiar with Python, we encourage you to play around with the NLP pipeline and customize your analyses for more insight into poem-type linguistic differences.

For now, let's move onto the analysis in R.

Analyzing the Poem Dataset in R

We'll start by loading the processed poem dataset and exploring the features we've derived using the code in Listing 9-1.

```
#load poem data and set session seed
mydata<-read.csv("PoemData.csv")
summary(mydata)
```

Listing 9-1: A script that reads in the processed poem data

Adverbs tend to be the least-represented features, accounting for only 0–10.5 percent of words used in any given poem. Nouns tend to be the most frequent words appearing in this set of poems, accounting for 12.5–53.5 percent of words in a given poem. Personal pronouns are rare, with more than a quarter of the poems not containing a personal pronoun (likely due to the haiku, which tend not to use them). Punctuation usage varies quite a bit, from no representation in a haiku to nearly half of a free-verse poem being composed of punctuation (a list poem of medical diagnoses at a hospital). Given this variation, it's likely we have good features to use in our analyses.

Let's set the seed for our analyses and visualize the parts-of-speech features with t-SNE, using the shape of each visualized point as a designation of the poem type; add the following code to Listing 9-1:

```
#grab relevant pieces of data
mydata1<-mydata[,-1]
haiku<-which(mydata1$Type=="haiku")
mydata1$Type[haiku]<-1
mydata1$Type[-haiku]<-2
set.seed(1)

#create and plot t-SNE projections of poem data
library(dimRed)
t1<-getDimRedData(embed(mydata1[,-1],"tSNE",ndim=2,perplexity=5))
plot(as.data.frame(t1),xlab="Coordinate 1",ylab="Coordinate 2",
main="Perplexity=5 t-SNE Results",pch=ifelse(mydata1$Type==1,1,2))
```

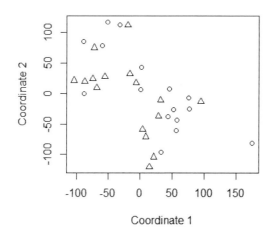

Figure 9-1: A t-SNE plot of poem features by poem type, with type represented by either circles or triangles (perplexity=5)

From the plot in Figure 9-1, we can see that our poems tend to separate out into clusters where most points are of the same type. This means a kernel-based model or nearest neighbor model is probably sufficient for classifying poem type by features.

Given some separation of points by features, it's likely we can apply algorithms to cluster our data and use supervised learning to understand which differences exist between haiku-type poems and other poems.

Let's divide our sample into training and test fractions and then apply a Euclidean-distance-based *k*-NN algorithm to classify poem types based on a poem's five nearest neighbors by adding to our code so far:

```
#create Canberra KNN models with different distances and five nearest
neighbors
library(knnGarden)
s<-sample(1:32,24)
train<-mydata1[s,]
test<-mydata1[-s,]
kc<-knnVCN(TrnX=train[,-1],OrigTrnG=train[,1],TstX=test[,-1],
K=5,method="euclidean")
```

Since our t-SNE plot (Figure 9-1) suggests some separation and points generally near poems of a similar type, it's likely that this model has worked well. Let's examine the predicted and true labels for our test set under the *k*-NN model through this addition to our code:

```
#examine performance
which(kc==test$Type)
> 1, 2, 3, 4, 5, 6, 7, 8
```

It looks like this model correctly classifies all test poems, suggesting a high-quality model that can separate types of poems based on features.

We'll examine these potential type differences by feature further using a 10-fold cross-validated dgLARS model in an additional step to our code:

```
#create a cross-validated dgLARS model
library(dglars)
dg1<-cvdglars(factor(Type)~.,data=mydata1,
family="binomial",control=list(nfold=10))
dg1
```

Because we are working with a small dataset, it's possible that one or more of your folds will have issues, giving a slightly different model than the results shown here:

```
#examine the dgLARS model
> dg1

Call:  cvdglars(formula=factor(Type) ~ .,family="binomial",data=mydata1,
    control=list(nfold=10))

Coefficients:
              Estimate
Int.          -1.4782
```

```
Punctuation_Length    0.3113
Adverb_Count         -0.2767
```

dispersion parameter for binomial family taken to be 1

```
Details:
   number of non-zero estimates: 3
      cross-validation deviance: 2.687
                            g: 0.4573
                       n. fold: 10
```

Your model should show differences in punctuation fraction and adverb fraction. Free-verse and formal-verse poems have higher rates of punctuation usage. Given that these types of poems are more likely to use full sentences rather than connected phrases, the differences in punctuation usage are consistent with expected differences.

The differences in adverb fractions aren't as expected. However, adverb usage is linked to many different transient and fixed personality traits. It's possible that haiku taps into a different transient mood or trait, creating style differences reflected in adverb usage. Subject matter may also influence this difference.

Importantly, this analysis shows that parts-of-speech and punctuation patterns vary within samples of writing by the same author in the same type of writing (poetry). Given that prior research suggests that some of these input features can be used to identify the likely author of a text, it may be prudent to rethink authorship prediction based on parts-of-speech analysis. Different subject matters, different types of writing, and different life stages of the author may influence word choice, sentence structure, and usage of punctuation.

To see how the poems group together within each poem type, let's visualize the persistence diagrams for each poem type through adding to our code:

```
#create the Vietoris-Rips complexes
library(TDAstats)
set1<-mydata1[mydata1$Type==1,-1]
set2<-mydata1[mydata1$Type==2,-1]

#compute Manhattan distance for set 1, compute Rips diagram, and plot
mm1<-dist(set1,"manhattan",diag=T,upper=T)
d1<-calculate_homology(as.matrix(mm1),dim=1,threshold=0,format="cloud")
plot_persist(d1)

#compute Manhattan distance for set 2, compute Rips diagram, and plot
mm2<-dist(set2,"manhattan",diag=T,upper=T)
d2<-calculate_homology(as.matrix(mm2),dim=1,threshold=0,format="cloud")
plot_persist(d2)
```

Figure 9-2 shows the persistence diagrams for haiku and non-haiku samples, highlighting some differences in poem clustering within type:

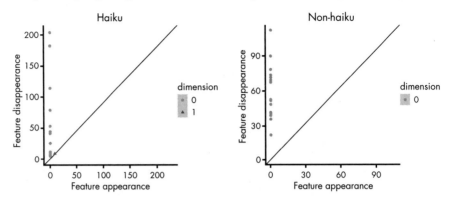

Figure 9-2: Persistence diagrams by poem type

As we can see in Figure 9-2, non-haiku poems tend to cluster more consistently, while haiku poems spread out without a lot of separation involving multiple poems grouped together. This suggests that haiku features vary from poem to poem, while non-haiku poems show more consistency in features. The spontaneity of the composition may result in a wider variety of poem structures found with haiku poems. Data related to spontaneousness of poem creation and time spent crafting the poem may shed light on conscious language usage differences between poems (such as haiku) that arise from a single moment and those that are created with more intent behind their creation.

Using Topology-Based NLP Tools

NLP as a field has evolved over the past years, and some common tools in NLP leverage topology to solve important problems. Early embedding tools tended to tally word frequencies within and across documents of interest to parse text data into numeric data that works well in machine learning algorithms. However, words often don't contain all the semantic information needed to make sense of a sentence or paragraph or entire document. Negatives, such as *no* or *not*, modify actions or actors within a sentence. For instance, "she did let him in the house" is a very different statement semantically than "she did not let him in the house." Depending on what you're trying to predict or classify, simple mappings from individual words to a matrix of numbers don't work well.

In addition, a document or collection of documents might have 30,000+ words that occur at least once. Most common words will occur often with little value added by their presence. Important words might occur only once or twice. This leads us into the problem of dimensionality again when we try to sift through the data for important trends.

Fortunately, recent years have seen some great developments in low-dimensional embeddings via topological mapping with special types of neural networks. *Pretrained transformer models* are neural networks that can pass information forward and backward through their hidden layers to obtain optimal topological mappings for text data. Pretrained transformer models learn low-dimensional embeddings of text data from massive training sets in the language or languages of interest. *BERT* (Bidirectional Encoder Representations from Transformers) and its sentence-based embedding cousin, *SBERT*, are two of the most common open source pretrained transformer models used to embed text data into lower-dimensional, dense matrices for use in machine learning tasks. BERT models can be extended to languages that are not currently supported, such as embeddings of text in Hausa or Lingala or Rushani; this has the potential to accelerate language translation services and preserve endangered languages. *GPT-3*, trained on a similar premise as BERT, has created accurate translations and chatbots that can parse meaning from input text rather than match keywords or eat up computing resources trying to process high-dimensional matrices within machine learning algorithms.

Using Python, we've built a BERT model on our poem set based on their serious tone or humorous tone to show how BERT embeddings can fit with our supervised learning tools in this book. (Note that the order of poems has changed from the original set with the data munging to get BERT embeddings and pass them back to a *.csv* file.) You can consult Python's transformer package to learn more about how BERT models are imported and leveraged in text embedding or refer to the Python scripts for this chapter (*https://nostarch.com/download/ShapeofData_PythonCode.zip*). However, we'll stick to importing the results into R and visualizing the data in a smaller dimension, as we did with our poem linguistic features earlier in the chapter. Let's create a t-SNE embedding and plot it with Listing 9-2.

```
#load poem data
#load data
mydata<-read.csv("BertSet.csv")

#create the embedding
library(dimRed)
t1<-getDimRedData(embed(mydata[,-1],"tSNE",ndim=2,perplexity=5))

#plot the results
plot(as.data.frame(t1),xlab="Coordinate 1",ylab="Coordinate 2",
main="Perplexity=5 t-SNE Results",pch=ifelse(mydata$Type=="serious",1,2))
```

Listing 9-2: A script that loads BERT data for serious and humorous poems, embeds the data with t-SNE, and plots the results

Figure 9-3 shows the embedding, which demonstrates that poems separate into clusters by poem type. This mirrors our haiku versus non-haiku poem results in Figure 9-1, where we saw features separating out by type in the t-SNE embedding. Again, this suggests that a machine learning classifier should work well with our dataset.

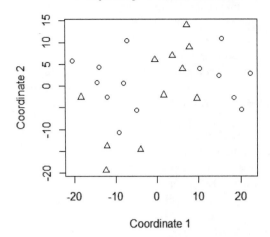

Figure 9-3: A plot of t-SNE embedding results, with serious or humorous types of poems denoted by circles and triangles, respectively (perplexity=5)

Now that we know a machine learning model may do well at classifying this data, let's fit a Lasso model with homotopy continuation to handle the small sample size by adding Listing 9-3:

```
#get training data split
set.seed(1)
s<-sample(1:25,0.85*25)
train<-mydata[s,]
test<-mydata[-s,]

#build Lasso model
library(lasso2)
etastart<-NULL
las1<-gl1ce(factor(Type)~.,train,family=binomial(link=probit),bound=5,standardize=F)
lpred1<-round(predict(las1,test,type="response"))

#examine results
lpred1
> 1 1 1 0 1
test$Type
> [1] serious serious serious humor serious
```

Listing 9-3: A script that loads BERT data for serious and humorous poems, splits the data into training and test sets, builds a Lasso model with homotopy continuation fitting, and shows prediction and actual values for test data

The dataset has 384 components from the embedded BERT model and 25 poems. Your version of R may split the data and fit the model differently, but in our version of R, the test data has three serious poems followed by a humorous one followed by a serious one. The model predicts a serious poem, a serious one, a serious one, a humorous one, and a serious one (giving a model accuracy of 100 percent). Given how small this training set is compared to the 384 predictors fed into our model, this is great performance. Note that the selected features have little meaning semantically, as they are simply embeddings. Coupling topology-based methods into pipelines to process and model small datasets can provide decent prediction where other models will fail entirely.

Summary

In this chapter, we applied several of the methods overviewed in the book on a linguistics question involving a dataset of features derived from NLP-processed poems; we also embedded our data and created a model to predict tone differences in our poem set.

For the first problem, we reduced the dimensionality of the dataset to visualize group differences, applied two supervised learning models to understand classification accuracy and important features distinguishing the poetry types, and visualized topological features that exist in both sets of poems. This showed us that language usage, particularly punctuation, varies across poem types.

We then looked at context-aware embeddings and predicting the tone of our poem set using a topology-based embedding method and a topology-based classification model, which showed that we could get fairly accurate prediction building a model from 384 embedded components and a training set of 21 poems.

In the next chapter, we'll wrap up the book by looking at ways to scale topological data analysis algorithms with distributed and quantum computing approaches.

10

MULTICORE AND QUANTUM COMPUTING

In the previous chapters, we overviewed many tools that come from the fields of geometry and topology. We saw how these algorithms can impact network analytics, natural language processing, supervised learning, time-series analytics, visualization, and unsupervised learning. Many more algorithms rooted in topology exist, including hybrid algorithms that improve existing models such as convolutional or recurrent neural networks, and the field has the potential to contribute thousands more to the field of machine learning.

However, one of the major issues facing the development and application of topological and geometric machine learning algorithms is computational cost. While most will not, calculating certain network metrics can scale to network sizes of one million or one billion (or more!). Other classes of algorithms, such as persistent homology or manifold learning, won't scale well; some will reach issues on a standard laptop at around 20,000 rows of data.

Calculating a distance matrix for a large set of data will also require a lot of time and computing resources.

Yet there are a couple of potential options for practitioners who hope to scale these methods. In this chapter, we'll review multicore approaches and quantum computing solutions. Solutions like the ones presented in this chapter are in their infancy, and as the algorithms in this book are adopted as big data algorithms, it's likely standard packages will be developed for quantum topological data analysis (TDA) algorithms and distributed TDA algorithms.

Multicore Approaches to Topological Data Analysis

One approach to algorithm scaling is to map pieces of the data to different computer cores or systems, compute the desired quantities on each core's worth of data at the same time, and reassemble each core's data quantity computations into a single dataset. For instance, suppose we want to calculate betweenness centrality for each vertex in a million-vertex network. Even if this is a sparse network without many edges, each betweenness centrality calculation will take a long time to compute. However, if we have access to a million cores, we can map the network to each core and compute betweenness centrality for one vertex in the network with each core. While the compute time may still be long, the total time needed to compute betweenness centrality for all one million vertices will be much less than what it would take computing each betweenness centrality one after another in a loop (as we did in prior chapters for much smaller networks). While a million cores isn't possible for most organizations, the time savings from 10 cores or 30 cores can be a substantial speedup.

It's also possible to map only part of a dataset to each core to compute some sort of metric. For instance, say we have a dataset of 300,000 individuals who completed an online customer survey. Calculating the distance between all 300,000 individuals' responses would take a lot of computing resources. However, distance matrices are necessary in many manifold learning and topological data analysis algorithms; it'd be good to have a quicker way to calculate this. We can map different pieces of the data to different cores to compute smaller distances matrices that can be assembled into the full 300,000-by-300,000-distance matrix after the cores have computed each piece. Again, we'll see large time savings.

In general, these multicore approaches fall under the umbrella of *distributed computing*, where multiple cores are leveraged to compute pieces of a problem to assemble when all cores have their solutions computed. Most cloud computing resources will have support for distributed computing, and some R and Python packages that can be run on the cloud support distributed computing options. However, it's usually up to the machine learning or data engineer to define distributed computing steps into the algorithm being used. Figure 10-1 shows a simple example of how data might be parsed and sent to different cores.

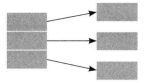

Figure 10-1: A mapping of three sections of data to three cores that will perform a mathematical or algorithmic step in the machine learning pipeline

In Figure 10-1, we parse our dataset into three pieces to map to three cores that will compute the quantities we desire. In a simple big data case, this might involve computing the minimum and maximum values found in each column of the dataset. We'd then compute the minimum and maximum values across cores, save that value, and spit out the minimum and maximum values for each column from the full set of data. It may not help much if we have a dataset of only a few million rows, but it will speed up the computational process a lot if we're dealing with trillions of rows of data.

Multicore approaches work well for network algorithms, such as those encountered in Chapters 2 through 4, and they have had some success with persistent homology and discrete exterior calculus. As long as the problem is local in nature or can be broken into smaller pieces for some steps, multicore approaches are viable. For instance, in robotics path-planning homotopy algorithms, we can break up the potential routes around obstacles and calculate some optimality for each route.

Unfortunately, few multicore versions of the algorithms in this book currently exist in R or Python. However, it is an area being actively explored in research, and if you're familiar with multicore frameworks on big data platforms, you are encouraged to play around with ways to apply this approach to persistent homology, manifold learning, or other geometry-based algorithms.

Quantum Computing Approaches

Another approach to scaling geometry-based algorithms is to implement them on a different type of computer that can leverage distributed computing natively. *Quantum computing* is a recent buzzword, and a lot of mystery and myths still surround the field, even within data science and software engineering. Progress has been faster than expected, but the technology is still in its early stages, with current systems having stark limitations and companies pouring money into hardware research and development. However, some question network algorithms already exist, and network science is one of the areas of machine learning that could benefit the most from quantum computing. In this section, we'll go through some basics of quantum computing and list some of the quantum algorithms related to graph theory and network science that exist as of 2023.

To start, quantum computing hardware can take on several different forms, each of which has its advantages and disadvantages. The type of circuit in the computer dictates what sort of algorithms can be developed on the machine, and some types of circuits are more amenable to network analytics. Two types of hardware seem to be the focus of most research and development these days, and they'll be the focus of this discussion.

The current systems have many limitations, including the need to cool the circuits, effects due to quantum scales (tunneling through energy barriers, fields created by interacting particles, and so on), and random error inherent in the qubits. Nevertheless, researchers have been able to explore potential new algorithms and quantum versions of existing algorithms through the quantum computers that exist and the simulation packages in Python. Graph theory and network algorithms, in particular, seem well suited to quantum computing, and the ability to search through combinatorics solutions simultaneously with qubits suggests that network science will get a boost when quantum computing scales to larger circuits.

Using the Qubit-Based Model

We'll start with the version of quantum hardware that is most like classical computers: the *qubit-based model*, which replaces classical bits with a quantum version of the bit, *qubits*, the quantum version of the 0 and 1 bits that underlie classical circuits. Rather than taking values of only 0 or 1 at a given time, qubits can exist simultaneously in the 1 and 0 state until the qubit is observed (where it will collapse to a 0 or a 1 state), and they can also rotate through computer gates to take fractional values. This means qubits can exist in many different states and be evolved through quantum gates into a final, optimized state. This makes them very flexible. Figure 10-2 shows the difference between bits and qubits.

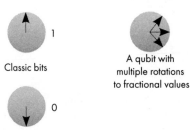

Figure 10-2: A plot comparing classical bits with qubits

Let's quickly go through the two main types of hardware that use qubits. Don't worry if you don't understand every term here; just keep the high-level ideas in mind. Two types of chipsets use qubits. Both rely on quantum principles of qubits (such as superposition) to speed up computation and improve accuracy. The first is a *gate-based circuit* that's similar to classical hardware, in which gate operations act on qubit states (such as the rotation gate). Gate-based circuits are the most common, and some examples include IBM's machines and Rigetti's machines; in general, gate-based algorithms speed up algorithms and provide easy benchmarking of algorithms. The other option, currently used by D-Wave on its machines, relies on quantum annealing (physical heating and cooling processes) through the changing of magnetic fields to act on qubit states rather than manipulating qubits through gates. In general, there aren't as many performance and accuracy guarantees or bounds on this type of system as a gate-based quantum hardware.

Using the Qumodes-Based Model

Besides the qubit model, the other dominant model is the *qumodes*-based circuit model, which is a photon-based circuit being developed by Xanadu, a Canadian company. Currently, software and algorithms based on photonic computing—which uses photons in place of qubits—are only simulations of the machine, but this type of simulation allows for the development of some interesting applications. This type of circuit employs wave functions, which are continuous distributions, in place of qubits (which collapse to a 1 or 0 when measured). Wave functions can then be squeezed, mapped, or operated on by other types of geometric transformations of the function without collapsing to a 1 or 0. This computer doesn't exist yet, but simulation programs do exist in Python (as of 2023), similar to simulation programs available for qubit circuits.

Using Quantum Network Algorithms

Several quantum network science algorithms exist that relate to properties of graphs. Clique-finding algorithms are particularly useful in network science, and quantum clique-finding algorithms already exist. Maximal clique algorithms seem to enjoy a speedup on the very small problems they've been tested on.

One important caveat of quantum algorithms is their probabilistic nature. Rather than getting, say, a list of cliques in the output, a quantum algorithm will run multiple times, creating arrays of clique lists, which can be combined into probabilistic scores of clique existence in the network. This can be helpful in prioritizing cliques for further parts of a project or zeroing in on the cliques of most interest within a very large network, though the latter will require much larger circuits than exist today.

Quantum maximum flow and minimum cut algorithms also exist; these algorithms aim to partition the graph into communities with the fewest possible cuts that maximize information flow on the graph. Applications thus far have explored importance-scoring uses to rank edges and vertices by importance to the graph structure and communication potential. They show some promise for sparse graphs and provide a probabilistic framework for deriving importance scores.

A basic quantum maximum flow and minimum cut algorithm using the R package QuantumOps does exist, though the capability is limited to small, sparse graphs. Using a qubits approach, the problem is first translated to a *quantum approximation optimization algorithm*, or *QAOA*. A QAOA formulation is a combinatorial algorithm that relies on the superposition of qubit states and something called *unitary operators* to solve optimization problems. Unitary operators are a type of linear algebra operator with special properties that match well to the underlying quantum mechanics of qubit circuits. Because of the probabilistic nature of solutions, it's best to run the algorithm multiple times for more exact answers. In this case, let's run the algorithm 10 times (an arbitrary number large enough to generate usable results with a probabilistic solution) and explore this function in a bit more depth by using the script in Listing 10-1, which reloads Graph 1 from Listing 4-1.

```
#load QuantumOps package
library(QuantumOps)

#transform Graph 1 from Listing 4-1 to unweighted graph
mydata1<-as.matrix(read.csv("Graph1w.csv",header=F))
mydata1[mydata1>0]<-1

#run the quantum approximation optimization algorithm maximum cut algorithm
q<-QAOA_maxcut(mydata1,p=10)

#explore the results
q
```

Listing 10-1: Running the maxcut algorithm on Graph 1, first seen in Chapter 4, 10 times

Note that the output from this algorithm contains 2^7 items, with our original graph containing six vertices. In general, this version of quantum maximum cut algorithms will include $2^{(n+1)}$ items in the output, with problems scaling to larger graphs. The output is also in raw format with this algorithm, leaving the user to translate the output to the most likely cuts made. A couple of existing Python packages give a more usable output, but this package provides a way for you to explore some qubit-based computing simulations within R.

In addition to maximum flow and minimum cut algorithms, there are quantum versions of random walk algorithms, and it would be possible to build a community-finding algorithm or PageRank-type algorithm from them. In fact, some quantum PageRank algorithms have been proposed in the past few years and studied theoretically. However, this type of application has not been explored much in practice thus far. It does provide an avenue for further research and potential applications to network science in the future.

Other examples that are more tied to graph theory than network science include graph-coloring algorithms and algorithms focused on testing graph properties (such as isomorphism and connectivity). Quantum querying on graphs and quantum versions of Dijkstra's algorithm or other shortest path algorithms are in their infancy, but they show promise in the future of quantum graph theory algorithms and quantum network science algorithms.

Though quantum computing is a nascent technology, graph theory and network science have already seen the potential gains from quantum algorithms, and it is likely more algorithms will be developed in the future. The extant algorithms generally run in Python for now, but it's likely that new interfaces to quantum computers will exist in the future, as more companies develop quantum computers and provide access to researchers.

Speeding Up Algorithms with Quantum Computing

Most algorithms do not have quantum computing or simulated quantum computing packages available yet. However, it's possible to leverage basic mathematical tools that do exist in quantum computing and simulated quantum computing packages to assemble algorithms such as the ones in this book step-by-step. The R package QuantumOps has some of these tools available for us to explore, so let's dive into an example of basic mathematics computations on a quantum system. Please note, this package only simulates algorithms that would be run on a quantum computer, so the speedups on your classical laptop won't be what you'd see on a real quantum computer. However, the following example will show one way that quantum algorithms are being developed to run on quantum computers to speed up computation for problems that require a lot of computational power as the complexity grows.

There are many algorithms that find the greatest common denominator between two numbers, and it's possible to do this within a quantum computing framework. The *greatest common denominator* refers to the largest number that will divide both numbers of interest; for instance, the largest number that divides both 12 and 20 is 4. We can factor 12 into its divisors (1, 2, 3, 4, 6, 12) and factor 20 into its divisors (1, 2, 4, 5, 10, 20). We can then see that 4 is the largest number to occur in both sets of divisors. This is a simple problem for us to do by hand when the numbers are small; however, in many encryption applications, the numbers to factor or determine to be prime can involve 12 or more digits. Factoring these requires a computer.

The gcd() function in QuantumOps finds the greatest common denominator between two numbers the way they would be found on a quantum computer. Let's try our example with 12 and 20 as our input numbers and see how this works with the gcd() function; see Listing 10-2.

```
#load QuantumOps package
library(QuantumOps)

gcd(12,20)
```

Listing 10-2: Finding the greatest common denominator of 12 and 20

You should see an output of 4 from Listing 10-2's code, which corresponds to the greatest common denominator of 12 and 20 that we found by hand. This function can compute common denominators of much larger numbers that would be impractical to compute by hand. Let's plug in two new, larger numbers (say, 14,267 and 11,345) and find their greatest common denominator with this function by modifying the gcd() function in Listing 10-2 to take these parameters:

```
gcd(14267,11345)
```

According to our output, the greatest common denominator between these two numbers is 1. Neither of these numbers is a prime, but they do not share any divisors. This algorithm runs quickly for numbers this large, and if you are interested in cybersecurity applications of factoring algorithms, you are encouraged to try numbers on the scale that you are using in your applications to further explore this function and its capabilities. On quantum systems, it's possible to run this algorithm for much larger numbers in a reasonable compute time. This means that security algorithms based on factoring will not perform well once quantum computing becomes more accessible outside of research settings.

The QuantumOps package does not include some of the more advanced mathematical tools upon which algorithms are built, but it's possible to define dot products, norms, and other linear algebra tools important to distance metric design, as well as define other linear algebra tools underlying common machine learning tasks. On a real quantum computer, we'd be able to run algorithms much more quickly and for much larger problems than the ones we've considered. However, the QuantumOps package allows us to explore some of what does exist in the quantum algorithm research that is done on quantum computers.

Using Image Classifiers on Quantum Computers

One of the hot topics in machine learning today is *image classification*, in which machine learning is used to classify input images according to the category labels it is given. Typically, you'll have a set of images with category labels associated to them that can be used to train a machine learning algorithm to identify characteristics (such as color, lines, circles, or more complicated patterns) that signal an image belongs to a certain class. For instance, consider the problem of labeling certain types of plants or animals based on pictures that may or may not contain one or more of the categories we're hoping to automatically tag. Consider Figure 10-3, which depicts the flowering part of a cannonball tree.

Figure 10-3: A picture of a blooming cannonball tree at Fairchild Tropical Gardens (native to Central and South America)

Now look at Figure 10-4, which depicts an elephant.

Figuro 10-4: A picture of an elephant walking down a road in Kruger National Park, South Africa

There are many challenges to classifying images such as these. In Figure 10-3, for instance, we don't see the entire tree in the image (making it difficult to recognize as a tree rather than a bush or vine), and the tree is in bloom (meaning that other pictures of a cannonball tree may not have flowers in them). In Figure 10-4, the elephant is walking away from the camera (meaning there is no animal face), the image has a lens filter applied (changing the natural color), and the image contains other types of things (plants, a road, and so on). Most real-world sets of images don't contain full images of only the category of interest that are plainly visible in the same color scheme (think of all the images of a cat that come up when you try searching Google Images for a cat). A good classifier needs to generalize to a lot of different types of cats in different lighting with other things included in the image.

We might also be facing category imbalance among the set of images we're using to train the algorithm. We might have a lot of pictures of orchids, tulips, jaguars, and kangaroos, but we may have relatively few pictures of black bat flowers, Gaboon vipers, or African dwarf sawsharks. For instance, we may have an entirely new plant that isn't found in nature or a rare plant or flower for which many images don't exist, such as the hybrid shown in Figure 10-5.

Figure 10-5: A picture of a genetically engineered new species of plant at Fairchild Tropical Gardens

Typically, image classification algorithms involve pretrained or custom-built convolutional neural networks, which were discussed in Chapter 1. To quickly review, convolutional neural networks (CNNs), a type of deep learning algorithm, find relevant features by optimizing maps from the input image layer to a series of different types of layers to the output layer containing the category labels. Within a CNN architecture, some layers find salient features within categories of images, which are pooled in other layers that find the best feature sets from prior layers and feed them into the next layer of feature-finding layers. It's common for these architectures to involve many layers, and they usually require a large set of training data to perform well. Further, they require a lot of computational power.

Because quantum computing offers faster runtimes for many algorithms and can leverage superposition to broaden search capabilities, the merging of deep learning algorithms such as CNNs with quantum computing offers a synergy to find relevant features more quickly and with less input data. In many real-world applications, we don't have access to every image that comes up on a Google search. We might have only a few hundred medical images of a rare condition or not have the computational power to optimize hundreds of parameters on an animal image set that includes thousands of pictures of hundreds of snake species that pose a threat to farmers or villages in the developing world.

However, many quantum neural network algorithms exist as of 2023, and many show competitive performance on problems such as image classification (though scaling is still an issue given the qubit number limitations of current quantum computers). We'll explore one recently developed in South Africa to handle image classification problems similarly to CNNs, the circuit-centric quantum classifier that exists in the QuantumOps package for general usage and usage on a famous image analytics benchmark dataset.

The details of circuit-centric quantum classifiers are a bit physics-heavy. It's okay if you don't follow all of this; we'll see it in action later in this section.

The basic approach of the circuit-centric quantum classifier is to learn the quantum gate parameters of the circuit through supervised learning, such as fitting hidden neural network–layer parameters in deep learning algorithms. The independent variables are coded into quantum systems amplitudes, which are manipulated by quantum gates. These single- and two-qubit gates are optimized through single-qubit measurement, which collapses the system from superposition into a single state. Gradients are learned by multiple runs of the algorithm, as in the approach in the quantum min cut and max flow algorithm earlier in this chapter. As with deep learning algorithms, we use dropout regularization fractions, which prune and add to the circuit during each iteration. The result is an optimized quantum architecture that modifies independent variable sets to create high-quality suggested dependent variable labels based on the training set independent and dependent variables.

While the exact details of how the quantum operators train the algorithm are beyond the scope of this book, the algorithm essentially modifies the quantum physics governing the processing of independent variables through the quantum circuit to produce accurate predictions of the dependent variable. One of the advantages of this approach is fewer parameters in the neural network architecture and training process to train compared to CNNs, which speeds up the fitting process and avoids the need for deep expertise in architecture design to obtain an architecture that fits the problem well.

The data we'll explore comes from one of the most common image analytics benchmark datasets. The MNIST dataset contains tens of thousands of images of handwritten digits (0–9). These were collected from sets of 250 and 500 writers and combined into a single dataset. Because different people and different cultures have different ways of writing numbers (such as the slashed 7 in certain parts of Europe to distinguish it from a 1) and different people often have different slants to their writing (right, left, center, down, up, straight, and so on), classifying which digit is in the image is a more challenging problem than it first appears.

Let's dive into an example that applies this circuit-centric quantum classifier in QuantumOps to the recognition of a single digit in the MNIST dataset with Listing 10-3.

```
#load QuantumOps package
library(QuantumOps)

QuantumMNIST256Classifier(matrix(sample(256,replace=TRUE),nrow=1),array(1),0,1,1,.001,1,"test")
```

Listing 10-3: *Training a quantum classifier to recognize the digit 0*

Listing 10-3 creates a sample of images with the target class set to the digit 0 (classifying 0 versus any other number), a learning parameter of 1, no decay of the learning rate over each iteration, a low bias (which allows other parameters to update faster), and one training iteration. Depending on your machine, it may take a while to run, as the algorithm is simulating a quantum system; it may even require a machine with more computational power. With a CNN, we'd need many more parameters associated with the network (not to mention needing to tune the architecture for the number and type of hidden layers), more training time, and more computational power. With this code, we optimize 33 quantum gates and find a matrix of the entire optimized classifier circuit. This represents the quantum architecture that will parse handwritten digits corresponding to 0 versus every other digit. We could repeat this process until we find quantum classifiers for each digit class in MNIST. On an actual quantum computer, this implementation would be much faster, as none of the circuitry would need to be simulated on a classical circuit.

While other incarnations of quantum neural networks exist, most don't have open source implementations yet. We encourage you to explore extant papers and tinker around with possible implementations of quantum neural networks on existing quantum systems. Most quantum computing companies have partnerships available for researchers in academia and industry to accelerate quantum algorithm design progress, and if you have access to a quantum computer, you can run the algorithms of this chapter (and other simulated packages) on a real quantum computer to leverage gains in computational speed.

Summary

In this chapter, we explored distributed computing solutions to scale algorithms, such as our computationally intensive network and TDA algorithms. We also introduced quantum computing frameworks, simulated a quantum graph algorithm for finding min flow and max cut solutions, ran a simulation of quantum greatest common denominator algorithms, and explored a simulated quantum classifier in the spirit of CNNs to explore the potential of quantum. As quantum computers grow in size and availability to researchers and industry data scientists, it's likely that you'll have access to quantum computers in the future to implement these types of algorithms on real systems that will improve performance over classical algorithms.

Given the current pace of circuit design, other solutions may be developed in the future to help scale the algorithms in this book that struggle on big data within current computational infrastructure, and we hope that tools to scale algorithms will help accelerate the development of new tools from the fields of geometry and topology. The field of topological data analysis is rapidly growing. More research in the field may lead to novel algorithms that solve general or niche problems in machine learning and data analysis.

REFERENCES

Chapter 1

Ahmed, Zo, Bertie Vidgen, and Scott A. Hale. "Tackling racial bias in automated online hate detection: Towards fair and accurate detection of hateful users with geometric deep learning." *EPJ Data Science* 11, no. 1 (2022): 8.

Alloghani, Mohamed, et al. "A systematic review on supervised and unsupervised machine learning algorithms for data science." *Supervised and Unsupervised Learning for Data Science* (2020): 3–21.

Draper, Norman R., and Harry Smith. "'Dummy' variables." *Applied Regression Analysis* (1998): 299–325.

Feuerriegel, Stefan, Mateusz Dolata, and Gerhard Schwabe. "Fair AI." *Business & Information Systems Engineering* 62, no. 4 (2020): 379–384.

Hridayami, Praba, I. Ketut Gede Darma Putra, and Kadek Suar Wibawa. "Fish species recognition using VGG16 deep convolutional neural network." *Journal of Computing Science and Engineering* 13, no. 3 (2019): 124–130.

Khattak, Faiza Khan, et al. "A survey of word embeddings for clinical text." *Journal of Biomedical Informatics* 100 (2019): 100057.

Köppen, Mario. "The curse of dimensionality." In *Proceedings of the 5th Online World Conference on Soft Computing in Industrial Applications (WSC5)*, 4–8. Vol. 1. World Federation on Soft Computing, 2000.

LeCun, Yann, Yoshua Bengio, and Geoffrey Hinton. "Deep learning." *Nature* 521, no. 7553 (2015): 436–444.

Mitchell, J. Clyde. "Social networks." *Annual Review of Anthropology* 3 (1974): 279–299.

Nelder, John Ashworth, and Robert W.M. Wedderburn. "Generalized linear models." *Journal of the Royal Statistical Society: Series A (General)* 135, no. 3 (1972): 370–384.

Orimoloye, Israel R., et al. "Geospatial analysis of wetland dynamics: Wetland depletion and biodiversity conservation of Isimangaliso Wetland, South Africa." *Journal of King Saud University - Science* 32, no. 1 (2020): 90–96.

Patrick, Edward A., and Frederic P. Fischer III. "A generalized k-nearest neighbor rule." *Information and Control* 16, no. 2 (1970): 128–152.

Rao, J., et al. "Bridging the gap between relevance matching and semantic matching for short text similarity modeling." In *Proceedings of the 2019 Conference on Empirical Methods in Natural Language Processing and the 9th International Joint Conference on Natural Language Processing (EMNLP-IJCNLP)*, edited by Sebastian Padó and Ruihong Huang, 5370–5381. Stroudsburg, PA: Association for Computational Linguistics, 2019.

Tumer, Kagan, and Joydeep Ghosh. "Analysis of decision boundaries in linearly combined neural classifiers." *Pattern Recognition* 29, no. 2 (1996): 341–348.

van der Aalst, Wil M.P., et al. "Process mining: A two-step approach to balance between underfitting and overfitting." *Software & Systems Modeling* 9, no. 1 (2010): 87–111.

Wang, Juanjuan, et al. "Classification of imbalanced data by using the SMOTE algorithm and locally linear embedding." In *2006 8th International Conference on Signal Processing*. Vol. 3. New York: Institute of Electrical and Electronics Engineers, 2006.

Chapter 2

Barabási, Albert-László, and Réka Albert. "Emergence of scaling in random networks." *Science* 286, no. 5439 (1999): 509–512.

Csardi, G., and Nepusz, T. (2006). "The igraph software package for complex network research." *InterJournal, Complex Systems*, no. 1695 (2006): 1–9.

Erdős, Paul, and Alfréd Rényi. "On the evolution of random graphs." *Publication of the Mathematical Institute of the Hungarian Academy of Sciences* 5, no. 1 (1960): 17–60.

Estrada, Ernesto, Franck Kalala-Mutombo, and Alba Valverde-Colmeiro. "Epidemic spreading in networks with nonrandom long-range interactions." *Physical Review E* 84, no. 3 (2011): 036110.

Fiedler, Miroslav. "Algebraic connectivity of graphs." *Czechoslovak Mathematical Journal* 23, no. 2 (1973): 298–305.

Fountsop, Arnauld Nzegha, Jean Louis Ebongue Kedieng Fendji, and Marcellin Atemkeng. "Deep learning models compression for agricultural plants." *Applied Sciences* 10, no. 19 (2020): 6866.

Giansiracusa, Noah, and Cameron Ricciardi. "Geometry in the courtroom." *The American Mathematical Monthly* 125, no. 10 (2018): 867–877.

Jalili, Mahdi, and Xinghuo Yu. "Enhancement of synchronizability in networks with community structure through adding efficient inter-community links." *IEEE Transactions on Network Science and Engineering* 3, no. 2 (2016): 106–116.

Leonard, Rosemary, Debbie Horsfall, and Kerrie Noonan. "Identifying changes in the support networks of end-of-life carers using social network analysis." *BMJ Supportive & Palliative Care* 5, no. 2 (2015): 153–159.

Li, Lun, et al. "Towards a theory of scale-free graphs: Definition, properties, and implications." *Internet Mathematics* 2, no. 4 (2005): 431–523.

Lou, Tiancheng, et al. "Learning to predict reciprocity and triadic closure in social networks." *ACM Transactions on Knowledge Discovery from Data (TKDD)* 7, no. 2 (2013): 1–25.

Moore, Christine, John Grewar, and Graeme S. Cumming. "Quantifying network resilience: Comparison before and after a major perturbation shows strengths and limitations of network metrics." *Journal of Applied Ecology* 53, no. 3 (2016): 636–645.

Page, Lawrence, et al. *The PageRank Citation Ranking: Bringing Order to the Web*. Technical report. Stanford InfoLab, 1999.

Saucan, Emil, et al. "Discrete curvatures and network analysis." *MATCH Communications in Mathematical and in Computer Chemistry* 80, no. 3 (2018).

Seshadhri, Comandur, Tamara G. Kolda, and Ali Pinar. "Community structure and scale-free collections of Erdős-Rényi graphs." *Physical Review E* 85, no. 5 (2012): 056109.

Sparrow, Malcolm K. "The application of network analysis to criminal intelligence: An assessment of the prospects." *Social Networks* 13, no. 3 (1991): 251–274.

Struweg, Ilse. "A Twitter social network analysis: The South African health insurance bill case." In *Conference on e-Business, e-Services and e-Society*, 120–132. Cham, Switzerland: Springer, Cham, 2020.

Valente, Thomas W. "Social network thresholds in the diffusion of innovations." *Social Networks* 18, no. 1 (1996): 69–89.

Wasserman, Stanley, and Katherine Faust. *Social Network Analysis: Methods and Applications.* Cambridge: Cambridge University Press, 1994.

Watts, Duncan J., and Steven H. Strogatz. "Collective dynamics of 'small-world' networks." *Nature* 393, no. 6684 (1998): 440–442.

Weber, Melanie, Emil Saucan, and Jürgen Jost. "Characterizing complex networks with Forman-Ricci curvature and associated geometric flows." *Journal of Complex Networks* 5, no. 4 (2017): 527–550.

Chapter 3

Abueldahab, Sheima M.E., and Franck Kalala Mutombo. "SIR model and HIV/AIDS in Khartoum." *Open Access Library Journal* 8, no. 4 (2021): 1–10.

Al Hasan, Mohammad, et al. "Link prediction using supervised learning." *SDM06: Workshop on Link Analysis, Counter-Terrorism and Security* 30 (2006): 798–805.

de Souza, Danillo Barros, et al. "Using discrete Ricci curvatures to infer COVID-19 epidemic network fragility and systemic risk." *Journal of Statistical Mechanics: Theory and Experiment* 2021, no. 5 (2021): 053501.

Emmons, Scott, et al. "Analysis of network clustering algorithms and cluster quality metrics at scale." *PLOS One* 11, no. 7 (2016): e0159161.

Estrada, Ernesto, Franck Kalala-Mutombo, and Alba Valverde-Colmeiro. "Epidemic spreading in networks with nonrandom long-range interactions." *Physical Review E* 84, no. 3 (2011): 036110.

Jamal, Wasifa, et al. "Classification of autism spectrum disorder using supervised learning of brain connectivity measures extracted from synchrostates." *Journal of Neural Engineering* 11, no. 4 (2014): 046019.

Kunegis, Jérôme. "Konect: The Koblenz Network Collection." In *Proceedings of the 22nd International Conference on World Wide Web*, 1343–1350. New York: Association for Computing Machinery, May 2013.

Lichtenwalter, Ryan N., Jake T. Lussier, and Nitesh V. Chawla. "New perspectives and methods in link prediction." In *Proceedings of the 16th ACM SIGKDD International Conference on Knowledge Discovery and Data Mining*, 243–252. New York: Association for Computing Machinery, July 2010.

Liu, Renming, et al. "Supervised learning is an accurate method for network-based gene classification." *Bioinformatics* 36, no. 11 (2020): 3457–3465.

Reichardt, Jörg, and Stefan Bornholdt. "Statistical mechanics of community detection." *Physical Review E* 74, no. 1 (2006): 016110.

Sanchez-Oro, Jesus, and Abraham Duarte. "Iterated Greedy algorithm for performing community detection in social networks." *Future Generation Computer Systems* 88 (2018): 785–791.

Sieranoja, Sami, and Pasi Fränti. "Adapting k-means for graph clustering." *Knowledge and Information Systems* 64, no. 1 (2022): 115–142.

Souabi, Sonia, et al. "A recommendation approach based on community detection and event correlation within social learning network." In *International Conference Europe Middle East & North Africa Information Systems and Technologies to Support Learning*, 65–74. Cham, Switzerland: Springer, Cham, 2019.

Yang, Zhao, René Algesheimer, and Claudio J. Tessone. "A comparative analysis of community detection algorithms on artificial networks." *Scientific Reports* 6, no. 1 (2016): 1–18.

Chapter 4

Akkoyunlu, Eralp Abdurrahim. "The enumeration of maximal cliques of large graphs." *SIAM Journal on Computing* 2, no. 1 (1973): 1–6.

Dika, Sandra L., and Kusum Singh. "Applications of social capital in educational literature: A critical synthesis." *Review of Educational Research* 72, no. 1 (2002): 31–60.

Edelsbrunner, Herbert, and Dmitriy Morozov. "Persistent homology: Theory and practice." In *European Congress of Mathematics, Kraków, 2–7 July, 2012*, 31–50. Helsinki, Finland: EMS Press, 2014.

Ghrist, Robert. "Barcodes: The persistent topology of data." *Bulletin of the American Mathematical Society* 45, no. 1 (2008): 61–75.

Hauser, Christoph, Gottfried Tappeiner, and Janette Walde. "The learning region: The impact of social capital and weak ties on innovation." *Regional Studies* 41, no. 1 (2007): 75–88.

Jonsson, Jakob. *Simplicial Complexes of Graphs*. Stockholm: Royal Institute of Technology, 2005.

Knill, Oliver. "On the dimension and Euler characteristic of random graphs." arXiv preprint arXiv:1112.5749 (2011).

Lee, Hyekyoung, et al. "Discriminative persistent homology of brain networks." In *2011 IEEE International Symposium on Biomedical Imaging: From Nano to Macro*, 841–844. New York: Institute of Electrical and Electronics Engineers, 2011.

Lee, Hyekyoung, et al. "Computing the shape of brain networks using graph filtration and Gromov-Hausdorff metric." In *International Conference on Medical Image Computing and Computer-Assisted Intervention*, 302–309. Berlin, Heidelberg: Springer, 2011.

Lee, Hyekyoung, et al. "Weighted functional brain network modeling via network filtration." In *NIPS Workshop on Algebraic Topology and Machine Learning*. Vol. 3. San Diego: Neural Information Processing Systems, 2012.

Lütgehetmann, Daniel, et al. "Computing persistent homology of directed flag complexes." *Algorithms* 13, no. 1 (2020): 19.

Nziku, Dina Modestus, and John Joseph Struthers. "Female entrepreneurship in Africa: Strength of weak ties in mitigating principal-agent problems." *Journal of Small Business and Enterprise Development* 25, no. 3 (2018): 349–367.

Weber, Melanie, Emil Saucan, and Jürgen Jost. "Can one see the shape of a network?" arXiv preprint arXiv:1608.07838 (2016).

Zomorodian, Afra, and Gunnar Carlsson. "Computing persistent homology." In *Proceedings of the Twentieth Annual Symposium on Computational Geometry*, 347–356. New York: Association for Computing Machinery, 2004.

Chapter 5

Alt, Helmut, and Michael Godau. "Computing the Fréchet distance between two polygonal curves." *International Journal of Computational Geometry & Applications* 5, no. 1, part 2 (1995): 75–91.

Anowar, Farzana, Samira Sadaoui, and Bassant Selim. "Conceptual and empirical comparison of dimensionality reduction algorithms (PCA, KPCA, LDA, MDS, SVD, LLE, ISOMAP, LE, ICA, t-SNE)." *Computer Science Review* 40 (2021): 100378.

Arowolo, Micheal Olaolu, et al. "Optimized hybrid investigative based dimensionality reduction methods for malaria vector using KNN classifier." *Journal of Big Data* 8, no. 1 (2021): 1–14.

Balasubramanian, Mukund, and Eric L. Schwartz. "The isomap algorithm and topological stability." *Science* 295, no. 5552 (2002): 7.

De Maesschalck, Roy, Delphine Jouan-Rimbaud, and Désiré L. Massart. "The mahalanobis distance." *Chemometrics and Intelligent Laboratory Systems* 50, no. 1 (2000): 1–18.

Diedrich, Holger, and Markus Abel. lle (Version 1.1). Package. March 21, 2012. *http://cran.nexr.com/web/packages/lle/index.html*.

Drost, Hajk-Georg. "Philentropy: Information theory and distance quantification with R." *Journal of Open Source Software* 3, no. 26 (2018): 765.

Faisal, M., and E.M. Zamzami. "Comparative analysis of inter-centroid K-Means performance using Euclidean distance, Canberra distance and Manhattan distance." *Journal of Physics: Conference Series* 1566, no. 1 (2020): 012112.

Ghorbani, Hamid. "Mahalanobis distance and its application for detecting multivariate outliers." *Facta Universitatis, Series Mathematics and Informatics* 34, no. 3 (2019): 583–595.

Gower, John Clifford. "Properties of Euclidean and non-Euclidean distance matrices." *Linear Algebra and Its Applications* 67 (1985): 81–97.

Higuchi, Tomoyuki. "Approach to an irregular time series on the basis of the fractal theory." *Physica D: Nonlinear Phenomena* 31, no. 2 (1988): 277–283.

Huttenlocher, Daniel P., Gregory A. Klanderman, and William J. Rucklidge. "Comparing images using the Hausdorff distance." *IEEE Transactions on Pattern Analysis and Machine Intelligence* 15, no. 9 (1993): 850–863.

Jongbo, Olayinka Ayodele, et al. "Development of an ensemble approach to chronic kidney disease diagnosis." *Scientific African* 8 (2020): e00456.

Kraemer, Guido, and Maintainer Guido Kraemer. dimred (Version 0.2.6). Package. July 11, 2022. *https://cran.r-project.org/web/packages/dimRed*.

Liebscher, Volkmar. "Gromov meets phylogenetics—new animals for the zoo of biocomputable metrics on tree space." arXiv preprint arXiv:1504.05795 (2015).

Mautner, F.I. "Geodesic flows on symmetric Riemann spaces." *Annals of Mathematics* (1957): 416–431.

Mémoli, Facundo. "Gromov-Hausdorff distances in Euclidean spaces." In *2008 IEEE Computer Society Conference on Computer Vision and Pattern Recognition Workshops*, 1–8. New York: Institute of Electrical and Electronics Engineers, 2008.

Mohibullah, Md, Md Zakir Hossain, and Mahmudul Hasan. "Comparison of Euclidean distance function and Manhattan distance function using k-medoids." *International Journal of Computer Science and Information Security* 13, no. 10 (2015): 61.

Mori, Usue, et al. TSdist (Version 3.7.1). Package. August 31, 2022. *https://cran.r-project.org/web/packages/TSdist*.

Oksanen, Jari, et al. vegan: Community Ecology Package (Version 2.6-4). October 11, 2022. *https://cran.r-project.org/web/packages/vegan*.

Rényi, Alfréd. "On the dimension and entropy of probability distributions." *Acta Mathematica Academiae Scientiarum Hungarica* 10, no. 1 (1959): 193–215.

Ringnér, Markus. "What is principal component analysis?." *Nature Biotechnology* 26, no. 3 (2008): 303–304.

Rodrigues, Érick O. "Combining Minkowski and Chebyshev: New distance proposal and survey of distance metrics using *k*-nearest neighbours classifier." *Pattern Recognition Letters* 110 (2018): 66–71.

Roweis, Sam T., and Lawrence K. Saul. "Nonlinear dimensionality reduction by locally linear embedding." *Science* 290, no. 5500 (2000): 2323–2326.

Sevcikova, Hana, et al. fractaldim (Version 0.8-5). Package. October 7, 2021. *https://cran.r-project.org/web/packages/fractaldim*.

Theiler, James. "Estimating fractal dimension." *Journal of the Optical Society of America A* 7, no. 6 (1990): 1055–1073.

Torgerson, Warren S. "Multidimensional scaling: I. Theory and method." *Psychometrika* 17, no. 4 (1952): 401–419.

Vallender, S.S. "Calculation of the Wasserstein distance between probability distributions on the line." *Theory of Probability & Its Applications* 18, no. 4 (1974): 784–786.

Van der Maaten, Laurens, and Geoffrey Hinton. "Visualizing data using t-SNE." *Journal of Machine Learning Research* 9, no. 11 (2008).

Wei, Boxian, et al. knnGarden (Version 1.0.1). Package. July 13, 2012. *https://cran.microsoft.com/snapshot/2019-03-14/web/packages/knnGarden*.

Wolpert, Scott A. "Geodesic length functions and the Nielsen problem." *Journal of Differential Geometry* 25, no. 2 (1987): 275–296.

Chapter 6

Augugliaro, Luigi, Angelo Mineo, and Ernst C. Wit. "dglars: An R package to estimate sparse generalized linear models." *Journal of Statistical Software* 59 (2014): 1–40.

Augugliaro, Luigi, Angelo Mineo, and Ernst C. Wit. "Differential geometric LARS via cyclic coordinate descent method." In *International Conference on Computational Statistics (COMPSTAT 2012)*, 67–79. The Hague, Netherlands: International Statistical Institute/International Association for Statistical Computing, 2012.

Babatunde, Seye, Richard Oloruntoba, and Kingsley Agho. "Healthcare commodities for emergencies in Africa: Review of logistics models, suggested model, and research agenda." *Journal of Humanitarian Logistics and Supply Chain Management* (2020).

Desbrun, Mathieu, et al. "Discrete exterior calculus." arXiv preprint math/0508341 (2005).

Gebhart, T., X. Fu, and R.J. Funk. "Go with the flow? A large-scale analysis of health care delivery networks in the United States using Hodge theory." In *2021 IEEE International Conference on Big Data (Big Data)*, 3812–3823. New York: Institute of Electrical and Electronics Engineers, 2021.

Hale, Trevor, and Christopher R. Moberg. "Improving supply chain disaster preparedness: A decision process for secure site location." *International Journal of Physical Distribution & Logistics Management* 35, no. 3 (2005): 195–207.

Hirani, Anil Nirmal. "Discrete exterior calculus." PhD thesis, California Institute of Technology, 2003.

Ngwenya, Ngonidzahe K., and Micheline J.A. Naude. "Supply chain management best practices: A case of humanitarian aid in southern Africa." *Journal of Transport and Supply Chain Management* 10, no. 1 (2016): 1–9.

Sommese, Andrew J., Jan Verschelde, and Charles W. Wampler. "Introduction to numerical algebraic geometry." In *Solving Polynomial Equations*, edited by Manuel Bronstein et al., 301–337. Berlin: Springer, 2005.

Xu, Qianqian, et al. "HodgeRank on random graphs for subjective video quality assessment." *IEEE Transactions on Multimedia* 14, no. 3 (2012): 844–857.

Yeh, I-Cheng, and Che-hui Lien. "The comparisons of data mining techniques for the predictive accuracy of probability of default of credit card clients." *Expert Systems with Applications* 36, no. 2 (2009): 2473–2480.

Chapter 7

Edelsbrunner, Herbert, and John Harer. "Persistent homology—a survey." *Contemporary Mathematics* 453 (2008): 257–282.

Farrelly, Colleen M., et al. "The analysis of bridging constructs with hierarchical clustering methods: An application to identity." *Journal of Research in Personality* 70 (2017): 93–106.

Gross, Miraca U.M. *Exceptionally Gifted Children*. New York: Routledge, 2002.

Pearson, P., D. Muellner, and G. Singh. "Tdamapper: Topological data analysis using mapper." 2015, *https://mran.microsoft.com/snapshot/2016-08-05/web/packages/TDAmapper/README.html*.

Singh, Gurjeet, Facundo Mémoli, and Gunnar E. Carlsson. "Topological methods for the analysis of high dimensional data sets and 3D object recognition." *PBG@ Eurographics* 2 (2007).

Wadhwa, Raoul R., et al. "TDAstats: R pipeline for computing persistent homology in topological data analysis." *Journal of Open Source Software* 3, no. 28 (2018): 860.

Chapter 8

Alexander, J.C., and James A. Yorke. "The homotopy continuation method: numerically implementable topological procedures." *Transactions of the American Mathematical Society* 242 (1978): 271–284.

Fendji, Jean Louis Ebongue Kedieng, et al. "Cost-effective placement of recharging stations in drone path planning for surveillance missions on large farms." *Symmetry* 12, no. 10 (2020): 1661.

Hernandez, Emili, Marc Carreras, and Pere Ridao. "A comparison of homotopic path planning algorithms for robotic applications." *Robotics and Autonomous Systems* 64 (2015): 44–58.

Lokhorst, Justin, et al. lasso2 (Version 1.2-19). Package. May 31, 2014. http://cran.nexr.com/web/packages/lasso2/index.html.

Tsukurimichi, Toshiaki, et al. "Conditional selective inference for robust regression and outlier detection using piecewise-linear homotopy continuation." *Annals of the Institute of Statistical Mathematics* (2022): 1–32.

Velez-Lopez, Gerardo C., et al. "A novel collision-free homotopy path planning for planar robotic arms." *Sensors* 22, no. 11 (2022): 4022.

Chapter 9

Abbott, Jade, and Laura Martinus. "Benchmarking neural machine translation for southern African languages." In *Proceedings of the 2019 Workshop on Widening NLP*, 98–101. Stroudsburg, PA: Association for Computational Linguistics, 2019.

Adelani, David Ifeoluwa, et al. "MasakhaNER: Named entity recognition for African languages." *Transactions of the Association for Computational Linguistics* 9 (2021): 1116–1131.

Agirrezabal, Manex, Inaki Alegria, and Mans Hulden. "Machine learning for metrical analysis of English poetry." In *Proceedings of COLING 2016, the 26th International Conference on Computational Linguistics: Technical Papers*, 772–781. Stroudsburg, PA: Association for Computational Linguistics, 2016.

Beyers, Chris. *A History of Free Verse*. Fayetteville: University of Arkansas Press, 2001.

Bird, Steven. "NLTK: The natural language toolkit." In *Proceedings of the COLING/ACL 2006 Interactive Presentation Sessions*, 69–72. Stroudsburg, PA: Association for Computational Linguistics, 2006.

Devine, Peter, Yun Sing Koh, and Kelly Blincoe. "Evaluating unsupervised text embeddings on software user feedback." In *2021 IEEE 29th International Requirements Engineering Conference Workshops (REW)*, 87–95. New York: Institute of Electrical and Electronics Engineers, 2021.

Devlin, Jacob, et al. "Bert: Pre-training of deep bidirectional transformers for language understanding." arXiv preprint arXiv:1810.04805 (2018).

Eiselen, Roald, and Martin Puttkammer. "Developing text resources for ten South African languages." In *Proceedings of the Ninth International Conference on Language Resources and Evaluation (LREC'14)*, 3698–3703. Luxembourg: European Language Resources Association, 2014

Jivani, Anjali Ganesh. "A comparative study of stemming algorithms." *International Journal of Computer Applications in Technology* 2, no. 6 (2011): 1930–1938.

Medhat, Walaa, Ahmed Hassan, and Hoda Korashy. "Sentiment analysis algorithms and applications: A survey." *Ain Shams Engineering Journal* 5, no. 4 (2014): 1093–1113.

Nadeau, David, and Satoshi Sekine. "A survey of named entity recognition and classification." *Lingvisticae Investigationes* 30, no. 1 (2007): 3–26.

Pennebaker, James W. "The secret life of pronouns." *New Scientist* 211, no. 2828 (2011): 42–45.

Plisson, Joël, Nada Lavrac, and Dunja Mladenic. "A rule based approach to word lemmatization." Ljubljana, Slovenia: Jožef Stefan Institute, 2004.

Reif, Emily, et al. "Visualizing and measuring the geometry of BERT." *Advances in Neural Information Processing Systems* 32 (2019).

Ross, Bruce, ed. *Journey to the Interior: American Versions of Haibun*. Clarendon, VT: Tuttle, 1998.

Visser, Ruan, and Marcel Dunaiski. "Sentiment and intent classification of in-text citations using BERT." In *Proceedings of 43rd Conference of the South African Institute of Computer Scientists and Information Technologists*, edited by Aurona Gerber, 129–145. Vol. 85. South African Institute of Computer Scientists and Information Technologists, 2022.

Webster, Jonathan J., and Chunyu Kit. "Tokenization as the initial phase in NLP." In *COLING 1992 Volume 4: The 14th International Conference on Computational Linguistics*. Stroudsburg, PA: Association for Computational Linguistics, 1992.

Yasuda, Kenneth. *Japanese Haiku: Its Essential Nature and History*. Clarendon, VT: Tuttle, 2011.

Chapter 10

Bauer, U., Kerber, M., and Reininghaus, J. "Distributed computation of persistent homology." In *2014 Proceedings of the Meeting on Algorithm Engineering and Experiments (ALENEX)*, 31–38. Philadelphia: Society for Industrial and Applied Mathematics, 2014.

Birman, Kenneth P. "The process group approach to reliable distributed computing." *Communications of the ACM* 36, no. 12 (1993): 37–53.

Cui, Shawn X., et al. "Quantum max-flow/min-cut." *Journal of Mathematical Physics* 57, no. 6 (2016): 062206.

Farhi, Edward, Jeffrey Goldstone, and Sam Gutmann. "A quantum approximate optimization algorithm." arXiv preprint arXiv:1411.4028 (2014).

Farrelly, Colleen M., and Uchenna Chukwu. "Benchmarking in quantum algorithms." *Digitale Welt* 3, no. 2 (2019): 38–41.

Fountsop, Arnauld Nzegha, Jean Louis Ebongue Kedieng Fendji, and Marcellin Atemkeng. "Deep learning models compression for agricultural plants." *Applied Sciences* 10, no. 19 (2020): 6866.

Fukushima, Kunihiko, and Sei Miyake. "Neocognitron: A self-organizing neural network model for a mechanism of visual pattern recognition." In *Competition and Cooperation in Neural Nets: Proceedings, Kyoto 1982*, edited by S. Amari and M.A. Arbib, 267–285. Berlin: Springer, 1982.

Markov, Igor L., and Mehdi Saeedi. "Faster quantum number factoring via circuit synthesis." *Physical Review A* 87, no. 1 (2013): 012310.

Mutanga, O., and A.K. Skidmore. "Integrating imaging spectroscopy and neural networks to map grass quality in the Kruger National Park, South Africa." *Remote Sensing of Environment* 90, no. 1 (2004): 104–115.

Resch, Salonik. QuantumOps (Version 3.0.1). Package. February 3, 2020. *https://cran.r-project.org/web/packages/QuantumOps*.

Schuld, Maria, et al. "Circuit-centric quantum classifiers." *Physical Review A* 101, no. 3 (2020): 032308.

Su, Xiaolong, et al. "Experimental preparation of eight-partite cluster state for photonic qumodes." *Optics Letters* 37, no. 24 (2012): 5178–5180.

Ubaru, Shashanka, et al. "Quantum topological data analysis with linear depth and exponential speedup." arXiv preprint arXiv:2108.02811 (2021).

Vidal, Guifre, and Christopher M. Dawson. "Universal quantum circuit for two-qubit transformations with three controlled-NOT gates." *Physical Review A* 69, no. 1 (2004): 010301.

Wittek, Peter. *Quantum Machine Learning: What Quantum Computing Means to Data Mining.* San Diego: Academic Press, 2014.

Yoon, Hee Rhang, and Robert Ghrist. "Persistence by parts: Multiscale feature detection via distributed persistent homology." arXiv preprint arXiv:2001.01623 (2020).

Chapter 6 Datasets

"For those of you with an IQ in the top 0.01% (profoundly gifted) what has your educational path looked like?" Quora. *https://www.quora.com/For-those-of-you-with-an-IQ-in-the-top-0-01-profoundly-gifted-what-has-your-educational-path-looked-like*.

Chapter 9 Dataset Poems

The following are the Farrelly poem credits by row number in the dataset:

1. *Frogpond*, volume 41:3
2. *Frogpond*, volume 41:3
3. *Presence*, Winter 2019
4. *Another Trip Around the Sun*, 2019
5. *Failed Haiku*, December 2020
6. *Bundled Wildflowers*, 2020
7. *Frogpond*, volume 43:3
8. *Failed Haiku*, January 2020
9. "Brother" in *Frogpond*, volume 42:1
10. "Writing on the Wall" in *Haibun Today*, volume 13:3, September 2019
11. "Fires and Stones" in *Haibun Today*, volume 12:4, December 2018
12. "Captain's Log, Major Azax" in *Leading Edge*, issue 75
13. "Snug Jackets" in *#Femku*, March 2019
14. "Browsing History" in *drifting-sands-haibun*, issue 1
15. "A Shadow in the Night" in *Night to Dawn*, issue 38
16. "Crossing the Styx" in *cattails*, April 2020
17. "Anya" in *Vine Leaves Literary Journal*, issue 10
18. "Overtown Store Front" in *StepAway*, 2013
19. "After War" in *Cacti Fur*, March 2017
20. "Jackson ER" in *POND*, volume 38
21. "Consuming Fire" in *Four and Twenty*, volume 4:11
22. *Four and Twenty*, volume 6:8
23. "Moonlit Night" in *The Marquette Journal*, Fall 2005
24. "Do You See Me Sleeping Over There" in *Lake City Lights*, 2012
25. "Reminder Notes in a Trash Can" in *The Casserole*, 2014
26. "Calvin Today" in *The Casserole*, 2014
27. "Slick City Streets" in *The MacGuffin*, volume 36:3
28. "Teachings of a Street Prophet" in *The Recusant*, 2012
29. "Blue Eyes" in *RiverLit*, 2013
30. "Ever After" in *The Transnational*, volume 1
31. "Saturday Night Fire" in *Vine Leaves Literary Journal*, issue 4
32. "Forgotten" in *Lake City Lights*, 2012

INDEX

A
adjacency matrices
    centrality, 27, 30, 33–35
    directed and undirected
        networks, 27
    disease spread tracking, 67, 69
    persistent homology, 91–92
    spectral theory, 49–50
    weighted networks, 27
Akaike information criterion (AIC), 137, 140
algebraic connectivity, 50, 51
alpha (attenuation parameter), 35, 39
alpha centrality. *See* Katz centrality
authority centrality
    defined, 35
        measuring in social networks, 39–41
averaged perceptron tagger, 183–184

B
Barabási-Albert model, 52
basis (Hamel basis), 133
BERT (Bidirectional Encoder
        Representations from
        Transformers), 189–190
beta, 70
Betti numbers, 85–86
    defined, 85
    Euler characteristic, 87
    examples of, 85–86
    persistent homology, 88–89
    subgroup mining, 156–157
    validating measurement tools, 161
betweenness of vertices
    applications of, 32–33
    bridges, 64
    community mining, 60
    disease spread tracking, 68
    graph filtration, 79
    measuring in social networks, 37, 38, 41, 42
    overview of, 32–33
    predictions with social media
        network metrics, 57, 59
    topological data analysis, 194
Bidirectional Encoder Representations
        from Transformers (BERT), 189–190
binomial distributions
    dispersion, 137
    entropy, 107–110
Bonacich centrality. *See* Katz centrality
bridges
    betweenness, 38, 64, 68
    disease spread tracking, 68
    predictions with social media
        network metrics, 56–57
    walktrap algorithm, 61
browseVignettes(), xxii

C
calculate_homology(), 157
Canberra distance, 101–102, 103, 118
Čech complexes, 82
centrality, 29–42
    applications of, 30
    applying clustering to social media
        dataset, 60
    authority centrality, 35
    betweenness of vertices, 32–33
    closeness of vertices, 31
    defined, 30
    degree of vertices, 30–31
    disease spread tracking, 68
    distance and, 24
    eigenvector centrality, 33–34

centrality *(continued)*
    graph filtration, 77–79
    hub centrality, 35
    Katz centrality, 35
    measuring in example social network, 36–42
    PageRank centrality, 34–35
    predictions with social media network metrics, 56–59
    spectral theory, 27, 49
    topological data analysis, 194
    topological dimension, 82
Chebyshev distance, 101, 116
choice ranking comparison, 149–152
    HodgeRank, 152
    missing information, 150–151
    no consistent preferences, 151
    overview of, 149–150
    preference loops, 150–151
circuit-centric quantum classifiers, 203–204
classification and classifiers
    bot account detection, 64, 66
    convolutional neural network classifiers, 110
    curse of dimensionality, 16–17
    decision boundaries, 10–11
    defined, 2
    homology, 85–86
    homotopy, 167, 169
    image classification, 18–20, 200–204
    logistic regression classifiers, 16–17
    metric geometry, 116–119
    overfitting and underfitting, 13
    overview of, 10–11
    poetry analysis project, 186, 189–190
    predicting edge formation, 59
    quantum classifiers, 203–204
    supervised classifiers, 59, 65
    support vector machine classifiers, 103
closed triangles, 43, 49
closeness of vertices
    measuring in social networks, 37, 38
    overview of, 31
CNNs (convolutional neural networks), 18–19, 110, 202–204

cohomology, 141, 146
community mining (clustering vertices), 59–64
    evaluating quality outcome of clusters, 61–62
    exploring networks with random walks, 61
    overview of, 59–60
    running clustering algorithms, 62–64
    spinglass clustering, 62
conditional Fisher information, 134–135
conditional Rao score, 135
connected components
    graph Laplacian, 50, 51
    homology, 85–86, 89
    random walk algorithms, 34
    subgroup mining, 156
convex optimization problems, 148
convolutional neural networks (CNNs), 18–19, 110, 202–204
COVID-19 pandemic, 72–73, 127
curl flow, 152
curse of dimensionality, 7, 14, 95
    geometric perspective, 17
    overview of, 13–17
    perturbed points in Euclidean space, 14–16

D

data geometry, 1–21
    machine learning, 2–4
        matching algorithms, 4
        supervised learning, 2–3
        unsupervised learning, 3
    structured data, 4–17
        dummy variables, 5–7
        numerical spreadsheets, 8–10
        supervised learning, 10–17
    unstructured data, 17–21
        image data, 18–20
        network data, 17–18
        text data, 20–21
data integrity, 4
data points
    in structured data, 4, 7–9
    supervised learning, 9, 11, 14, 17
    unsupervised learning, 3

data science geometry, 95–129
- distance metrics, 96–116
  - entropy, 107–110
  - norm-based distance metrics, 99–105
  - shape comparison, 110–116
  - small dataset simulation, 98–99
  - Wasserstein distance, 105–107
- fractals, 125–129
- k-nearest neighbors with metric geometry, 116–119
- manifold learning, 119–125
  - Isomap, 121–122
  - locally linear embedding, 122–124
  - multidimensional scaling, 120–122
  - t-distributed stochastic neighbor embedding, 124–125

decision boundaries
- classification, 10–11
- overfitting, 13

decision trees
- decision boundaries, 10, 11
- overfitting, 13

deep learning
- convolutional neural networks, 18, 202–203
- defined, 4
- geometric, 18
- Riemannian manifolds, 18
- vector embeddings, 20–21

degree of networks, 48–49

degree of vertices (degree centrality). *See also* Katz centrality
- applications of, 31
- community mining, 60
- Forman–Ricci curvature, 46
- graph filtration, 76, 78–79
- graph Laplacian, 50
- in-degree and out-degree, 30
- k-means clustering, 60, 61
- limitations of, 31
- measuring in social networks, 41
- overview of, 30–31
- scale-free graphs, 52

topological dimension, 82–84
triadic closure, 43

dendrograms, 89, 107, 158–160

density of networks
- disease spread tracking, 69, 71
- graph filtration, 77
- overview of, 48–49

dependent variables
- defined, 2
- dummy variables, 5
- image classification, 203
- link functions, 135
- regression, 11, 12
- supervised learning, 2–3
- vertex centrality metrics, 56, 58

dgLARS algorithm, 133–140
- credit default prediction, 138–140
- cross-validated vs. non-cross-validated, 136–140
- depression prediction, 136–138
- overview of, 133–136
- poetry analysis project, 186
- risk propensity measurement, 134–135

dglars package, 136

diameter of networks
- graph filtration, 79–80
- network comparison, 65–66
- overview of, 49

differential geometry, 88. *See also* dgLARS algorithm

differential geometry least angle regression algorithm. *See* dgLARS algorithm

Dijkstra's algorithm, 199

dimensionality
- curse of, 7, 13–17, 95
- defined, 14
- reduction of, 95, 119–120, 184
- unsupervised learning, 3

directed networks, 19
- applications of, 26
- authority centrality, 39–40
- converting undirected to, 39–40
- defined, 26
- degree of vertices, 30
- edges, 28

directed networks *(continued)*
    eigenvector centrality, 34
    hub centrality, 39–40
    interconnectivity of networks, 48
    networks in R, 26–27
    PageRank algorithm, 33
    Twitter, 17, 26
disaster logistics planning, 142–146
discrete exterior derivatives, 140–146
    cohomology, 141, 146
    differential forms, 141
    disaster logistics planning, 142–146
    engineering problems, 146
    overview of, 140, 141
    social network analysis, 141–142
dist( ), 99, 101, 104
distance metrics, 96–116
    entropy, 107–110
    norm-based distance metrics, 99–105
    overview of, 96–98
    shape comparison, 110–116
    small dataset simulation, 98–99
    Wasserstein distance, 105–107
distributed computing, 194–195
diversity of vertices, 42
Dow Jones Industrial Average (DJIA), 127–128
dummy variables, 5–7
    categorical variables, 5–6
    geometry of, 5–7
    multicollinearity, 7
D-Wave, 197

E

Ebola outbreak, 29
eccentricity of vertices
    diameter and, 49
    graph filtration, 80
    overview of, 45
    radius and, 49
edge lists, 26–27
edges
    adjacent, 28
    closeness of vertices, 31
    degree of vertices, 30
    density of networks, 48–49

    depiction of, 25
    directed and undirected networks, 17, 26
    disease spread tracking, 67–69, 70
    diversity of vertices, 42
    Erdös-Renyi graphs, 51–52
    Euler characteristic, 87
    Forman–Ricci curvature, 46–47
    graph filtration, 76–78, 80
    intracommunity and intercommunity edges, 61
    link prediction in social media, 58–59
    network comparison, 65
    overview of, 25
    path length, 28
    weighted and unweighted networks, 28–29
efficiency of networks, 49
efficiency of vertices, 44–45
eigen( ), 50
eigenvalues, 33, 49–50
eigenvectors, 33, 49–50
    eigenvector centrality, 33–39
        authority and hubness, 35
        Katz centrality and, 36
        measuring in social networks, 38–39
        overview of, 33–34
        PageRank centrality and, 34–35
elastic net regression, 101
EM algorithm, 171
entities
    named entity recognition, 180
    spread of, 66–68
    vertices and edges, 25
entropy, 107–110
    diversity of vertices, 42
    relative, 135
    Shannon entropy, 42
epidemiology
    centrality, 30
    disease spread tracking, 67–74
    spectral radius, 50
Erdös-Renyi graphs, 51–52
    network comparison, 65–66
    persistent homology, 90–93

Euclidean distance
    curse of dimensionality, 14, 16
    k-nearest neighbors, 118–119, 186
    multidimensional scaling, 121, 122
    network distance and, 29
    norm-based distance metrics, 99–103
    spreadsheet geometry, 9
Euclidean vector space
    curse of dimensionality, 15, 16
    defined, 8
    manifolds, 119
    multidimensional scaling, 120–122
    shape comparison, 113–116
    spreadsheet geometry, 8
    tangent spaces, 133
    vector embeddings, 21
Euler characteristic, 87–88
    Betti numbers, 87
    Gauss-Bonnet theorem, 88
    maximal cliques, 87
    negative, 87
    simplicial complexes, 87
expectation-maximization (EM) algorithm, 172

F

Facebook
    bot account detection, 18, 24
    degree of vertices, 30
    global network metrics, 47
    link prediction, 58
    network distance, 24
    text search, 20
    undirected networks, 17, 26
fast greedy clustering, 61–64
feature importance, 3
filtration
    graph filtration, 76–81
    network filtration, 75–94
Fisher information, 134–135
flag complexes, 82–83
fMRI. *See* functional magnetic resonance imaging
Forman–Ricci curvature
    differential geometry, 88
    disrupting communication and disease spread, 72–73
    overview of, 45–47
    stock market change point detection, 129
Forman–Ricci flow, 73–74
fractals, 125–129
Fréchet distance, 111–112
functional magnetic resonance imaging
    network comparison, 64, 66
    persistent homology, 90–93

G

gamma, 70
gate-based circuits, 196–197
Gauss-Bonnet theorem, 88
Gaussian distribution, 172–173
Gaussian noise, 14
gcd( ), 199–200
genomics, 17, 88, 136
    datasets, 86, 101, 119
geodesics, 33, 44, 121
    tangent spaces and, 96–97
geometric deep learning, 18
geospatial data, 8–9, 9
gerrymandering, 24
global network metrics, 47–51
    graph filtration, 79
    interconnectivity of networks, 48–49
    network comparison, 93
    spectral measures of networks, 49–51
    spreading processes on networks, 49
Google
    image search, 202–203
    PageRank algorithm, 33
    PageRank centrality, 34–35
    text search, 20
GPT-3, 189
gradient descent, 169–171
gradient flow, 152
graph diameter, 79–80
graph filtration, 76–81
    brain imaging studies, 80
    degree centrality, 78–79
    graph diameter, 79–80
graph Laplacian, 50–51
graph theory, 24, 195, 198–199

greatest common denominator, 199–200
greedy algorithms, 61–64
Gromov-Hausdorff distance, 113–116, 160
gromovlab package, 114

H

Hamel basis, 133
Hamming distance, 163–164
harmonic flow, 152
Hausdorff distance, 113, 160
hclust(), 156
heatmaps, 11, 89
help(), xxii
hierarchical clustering, 3, 89, 156–158, 163
Hodge-Helmholtz decomposition, 152
HodgeRank, 152
homology, 85–94
    Betti numbers, 85–86
    cohomology, 140–141, 146
    defined, 85
    differential geometry, 88
    Euler characteristic, 87–88
    persistent homology, 88–89, 129, 159, 195
        measurement validation and, 160–161
        network comparison and, 89–94, 155–156
        subgroup mining and, 156–157, 159, 162
homotopic Fréchet distance, 111
homotopy algorithms, 167–177
    comparing, 173
    homotopic, defined, 167
    homotopy-based regression, 169–174
    logistic regression vs. homotopy-based regression, 174–176
    overview of, 167–168, 169
hub centrality
    community mining, 60
    defined, 35
    graph filtration, 77
    measuring in social networks, 39–42
    unsupervised learning, 60
hyperparameters
    classification, 11
    defined, 3
    overfitting, 13
    regression, 12

I

IBM, 197
igraph library, 27, 29–30, 35, 43, 52, 85, 87, 90, 165
    cluster_edge_betweenness(), 64
    default value, 39
    eccentricity(), 49
    edge_density(), 48, 69
    efficiency(), 49
    sample_gnp(), 51
    sample_pa(), 52
    sample_smallworld(), 52
    sir(), 70
    spectrum(), 50
    transitivity(), 49
image classification
    convolutional neural networks, 18–20
    quantum computing approaches, 200–204
image data
    convolutional neural networks, 18–19, 20
    Forman–Ricci flow, 73–74
    overview of, 18–20
    persistent homology, 88–89
in-degree, 30
independent variables
    decision trees, 11
    defined, 2
    dgLARS algorithm, 135
    dimensionality, 14, 101
    dummy variables, 5, 7
    geometric deep learning, 18
    image classification, 19, 203
    multicollinearity, 7
    supervised learning, 2–3
    unsupervised learning, 3
    vertex centrality metrics, 57–58
instances. *See* data points
interconnectivity of networks, 40, 44, 47–49, 52

inverse Hamming distance, 164
Isomap, 121–122
isometric embedding, 113, 116

K

Katz centrality
    eigenvector centrality and, 35
    measuring in social networks, 39
    overview of, 35
k-means clustering, 3
    community mining, 59–60
    vs. Mapper algorithm, 161, 163
k-nearest neighbors (k-NN), 2–3
    decision boundaries, 10, 11
    dummy variables, 7
    metric geometry, 116–119
    overfitting, 13
    poetry analysis project, 186
    regression, 11–12
knnGarden package, 117
KONECT Windsurfer Network, 69–71
Kullback-Leibler divergence
    dgLARS algorithm, 135
    entropy, 108–110
    t-distributed stochastic neighbor embedding, 124

L

Lasso algorithm
    homotopy-based optimization, 172, 174–176
    Lasso regression, 101
    poetry analysis project, 190
lasso2 package, 172
linear dependence, 7
linear regression, 2–3
    dgLARS algorithm, 136, 138
    making predictions with social media network metrics, 57
    multicollinearity, 7
    supervised regression, 12
    vs. homotopy-based regression, 173–174, 176
link functions, 136
link prediction, 58–59
locally linear embedding (LLE), 122–124
local optima, 169–172, 174, 176

logistic regression, 2
    curse of dimensionality, 16
    decision boundaries, 10, 11
    dgLARS algorithm, 140
    link functions, 136
    multicollinearity, 7
    overfitting, 13
    vs. homotopy-based regression, 174–176
Louvain clustering, 62–64

M

machine learning categories, 2–4
    matching algorithms, 4
    supervised learning, 2–3
    unsupervised learning, 3
mahalanobis(), 103
Mahalanobis distance, 103–104, 105
Manhattan distance, 100–102
    k-nearest neighbors, 116, 118–119
    multidimensional scaling, 121, 122
    subgroup mining, 156–157
manifold hypothesis, 95
manifold learning, 119–125
    Isomap, 121–122
    locally linear embedding, 122–124
    multidimensional scaling, 120–122
    vs. principal component analysis, 119
    t-distributed stochastic neighbor embedding, 124–125
manifolds
    defined, 8
    distance metrics, 96, 98, 194
    Gauss-Bonnet theorem, 88
    homology, 85, 88
    Riemannian manifolds, 18
    tangent spaces, 132–133
Mapper algorithm, 161–166
    stepping through, 162–163
    using TDAmapper to find cluster structures in data, 163–166
matching algorithms, 4
Matlab, 152
maximal cliques, 82–84
    disaster logistics planning, 144
    Euler characteristic, 87
    quantum network algorithms, 197

MDS (multidimensional scaling), 120–122
median(), 70
metric geometry, 98
    fractals, 125–129
    k-nearest neighbors, 116–119
    manifold learning, 119–125
Minkowski distance, 101–102, 121
MNIST dataset, 204
model fit
    dgLARS algorithm, 137–138
    homotopy-based regression, 172
    nonlinearity, 147, 149
modularity, 61–64
Morse functions, 162
multicollinearity, 7, 133
multicore approaches, 193–195
multidimensional scaling (MDS), 120–122

N

named entity recognition, 180
natural language processing
    pipelines, 180–181
    topology-based tools, 188–191
network analysis, 55–74
    spread analysis, 66–74
        disease spread tracking between towns, 67–69
        disease spread tracking between windsurfers, 69–72
        disrupting communication and disease spread, 72–74
    supervised learning, 56–59
        diary entry prediction in social media, 56–58
        link prediction in social media, 58–59
    unsupervised learning, 59–64
        applying clustering to the social media dataset, 59–60
        community mining, 61–64
network comparison, 64–66
network depth, 31, 33
network distance, 28–29
    applications of, 24, 28–29
    defined, 28
    link prediction in social media, 59
    persistent homology, 91–93
    weighted and unweighted networks, 28
network filtration, 75–94
    graph filtration, 76–81
    homology, 85–94
        Betti numbers, 85–86
        Euler characteristic, 87–88
        persistent homology, 88–89
        comparison with, 90–94
    simplicial complexes, 81–85
network geometry, 23–54
    directed and undirected networks, 18
    global network metrics, 47–51
        interconnectivity of a network, 48–49
        spectral measures of a network, 49–51
        spreading processes on a network, 49
    models for real-world behavior, 51–53
        Erdös-Renyi graphs, 51–52
        scale-free graphs, 51–52
        Watts-Strogatz graphs, 52–53
network science, 24–25
network theory, 25–29
    directed networks, 26
    networks in R, 26–27
    paths and network distance, 28–29
    overview of, 17–18
    Riemannian manifolds, 18
    vertex metrics, 30–47
        centrality, 30–42
        diversity of vertices, 42
        eccentricity of vertices, 45
        efficiency of vertices, 44–45
        Forman–Ricci curvature, 45–47
        triadic closure, 43–44
networks, defined, 8
neural networks, 2
    convolutional, 18–19, 202–204
    decision boundaries, 10, 11, 13
    homotopy-based optimization, 172

quantum, 203–204
topology-based NLP tools, 189
neuroscience and brain imaging
graph filtration, 80
network comparison, 64
persistent homology, 90–93
NLP. *See* natural language processing
NLTK toolkit, 181, 183–184
nodes. *See* vertices
nonconvex objects, 147–148
nonlinear algebra, 146–149
convex optimization problems, 148
nonconvex objects, 147–148
numerical algebraic geometry, 147–149
vs. linear algebra, 146–147
nonlinear spaces, 132–140
dgLARS algorithm, 133–140
credit default prediction, 138–140
depression prediction, 136–138
overview of, 133–136
tangent spaces, 132–133, 135
nonselected predictors, 135
norms, defined, 99
numerical algebraic geometry, 147–149
numerical spreadsheets. *See* spreadsheets

O

observations. *See* data points
open triangles, 43
out-degree, 30
outlier detection, 24
network comparison, 66
stealth outliers, 104
subgroup mining, 159
overfitting
curse of dimensionality, 14, 16–17
overview of, 13

P

PageRank algorithm, 33, 198
PageRank centrality, 34–35
community mining, 60
link prediction in social media, 59
measuring in social networks, 38–39, 41, 42
paths, defined, 28
PCA (principal component analysis), 3, 119
perplexity, 124–125
Perron-Frobenius theorem, 34
persistence diagrams
measurement tool validation, 159–161
persistent homology, 89, 91–93
poetry analysis project, 187–188
shape comparison, 110
subgroup mining, 157
persistent homology, 88, 155–159
measurement tool validation, 160–162
multicore approaches, 195
network comparison, 90–94
outlier detection, 159
overview of, 88–89
stock market change point detection, 129
subgroup mining, 156–159
PET (positron emission tomography), 64, 90–93
philentropy package, 108
plot_persist(), 157
poetry analysis project, 180–188
analysis in R, 184–188
forms of poetry, 181–182
natural language processing pipeline, 180–181
normalizing vectors, 184
tagging parts of speech, 183–184
tokenizing text data, 183
topology-based NLP tools, 188
point clouds, 82, 88–89, 146, 156–157, 162–163
Poisson distribution, 56–57
Poisson regression, 57, 58
polynomials, 147–148
positron emission tomography, 64, 90–93
predictors. *See* independent variables
pretrained transformer models, 189
principal component analysis, 3, 119

probability density functions,
106–108, 109
probability distributions
entropy, 107–110
t-distributed stochastic neighbor
embedding, 124
Wasserstein distance, 105–107
propagation analysis. *See* spread
analysis
Pythagorean distance, 100. *See also*
Euclidean distance
Python
BERT model, 189
distributed computing, 194
help resources for, xxiii
natural language processing,
180–181
poetry analysis project, 183–184
quantum computing, 196–199

Q

quantum algorithms, 197–200, 204
quantum annealing, 197
quantum approximation optimization
algorithms (QAOA), 198
quantum classifiers, 203–204
quantum computing approaches,
193–205
image classifiers, 200–204
quantum algorithm development,
199–200
quantum network algorithms,
197–199
qubit-based model, 196–197
qumodes-based model, 197
quantum maximum flow and minimum
cut algorithms, 197–198
QuantumOps package, 198–200,
203–204
qubits
circuit-centric quantum
classifiers, 203
quantum network algorithms, 198
qubit-based model, 196–197
qumodes-based model, 197
Quora dataset, 136, 156, 160, 163,
166, 174

R

R (programming language)
downloading, xxi–xxii
help resources for, xxii
installing, xxii
installing packages, xxii
vignettes, xxii–xxiii
radius of networks, 35, 49–50
random forests, 2
decision boundaries, 11
regression, 11–12
random walk algorithms
diversity of vertices, 42
eigenvector centrality, 34, 38
link prediction, 59
map networks, 24
PageRank centrality, 35, 38–39
quantum algorithms, 198
walktrap algorithm, 61–64
redundant predictors, 133, 135
regex tokenizer, 183
regression
defined, 2
dummy variables, 5–6
elastic net regression, 101
homotopy-based regression,
169–176
journal ranking, 66
linear regression, 2–3, 12
dgLARS algorithm, 136, 138
making predictions with
social media network
metrics, 57
multicollinearity, 7
vs. homotopy-based
regression, 173–174, 176
logistic regression, 2
curse of dimensionality, 16
decision boundaries, 10, 11
dgLARS algorithm, 140
link functions, 136
multicollinearity, 7
overfitting, 13
vs. homotopy-based
regression, 174–176
overview of, 11–12
Poisson regression, 57, 58

reinforcement learning, 4
Ricci curvature, 45. *See also*
    Forman–Ricci curvature
Riemannian manifolds, 18
Rigetti, 197
RStudio, xxii

S

SBERT (sentence-based Bidirectional Encoder Representations from Transformers), 189
scale-free graphs, 52–53
    network comparison, 65–66
    persistent homology, 90–93
scatterplots, 11, 100, 119, 171
selected predictors, 135
Shannon entropy, 42
shape comparison, 110–116
    Fréchet distance, 111–112
    Gromov-Hausdorff distance, 113–116
    Hausdorff distance, 113
simplicial complexes, 81–85
    Čech complexes, 82
    cochains, 141
    Euler characteristic, 87
    filtering, 82
    flag complexes, 82–83
    Mapper algorithm, 162
    maximal cliques, 82–84
    overview of, 81–82, 84
    persistent homology, 88–89, 156–157
    topological dimension, 82–84
    Vietoris-Rips complexes, 82
simulated annealing, 62
single-linkage hierarchical clustering, 89, 156, 163
SIR model. *See* susceptible-infected-resistant model
SMOTE (Synthetic Minority Oversampling Technique), 8
social networks
    algebraic connectivity, 50
    bot account detection, 18, 24, 64, 66
    centrality, 30–42

clustering vertices, 59–64
diary entry prediction, 56–58
directed and undirected networks, 17–18
discrete exterior derivatives, 141–142
Forman–Ricci curvature, 46, 47
graph filtration, 76–80
influencers, 24, 30–31
link prediction, 58–59
network depth, 33
simplicial complexes, 82–83, 84
spread of misinformation, 9, 66–67
subgroup mining, 156–157
transitivity, 43, 44
triadic closure, 43
vertex vs. global metrics, 47
Watts-Strogatz graphs, 52
spectral gap of networks, 50
spectral measures of networks, 49–51
spectral radius, 35, 49–50, 79
spinglass algorithms, 62
spinglass clustering, 62–64
spread analysis, 66–74
    disease spread tracking
        between towns, 67–69
        between windsurfers, 69–72
    disrupting communication and disease spread, 72–74
spreading processes on networks, 49
spreadsheets
    adjacency matrices, 27
    classification, 10
    dummy variables, 5–7
    Euclidean vector space, 8
    geometry of, 8–10
    geospatial data, 8–9
    structured data, 4
stats package, 99
stock market change point detection, 127–129
strength (weighted degree) of vertices, 30, 82
structured data, 4–17
    defined, 4
    dummy variables, 5–7
    numerical spreadsheets, 8–10

Index **231**

structured data *(continued)*
    supervised learning, 10–17
        classification, 10–11
        curse of dimensionality, 14–17
        overfitting, 13
        regression, 11–12
subgroup mining
    Mapper algorithm, 161–166
    persistent homology, 156–159
supervised learning, 10–17
    algorithm viewed as function, 2–3
    classification, 10–11
    combining with unsupervised learning, 3
    curse of dimensionality, 14–17
    overfitting, 13
    overview of, 2–3
    prediction
        diary entry prediction in social media, 56–58
        link prediction in social media, 58–59
    regression, 11–12
    training and testing data, 3
support vector machines, 2, 172
susceptible-infected-resistant model
    defined, 68
    disease spread tracking
        between towns, 67–69
        windsurfers, 69–72
    disrupting communication and disease spread, 72–74
Synthetic Minority Oversampling Technique (SMOTE), 8

T

tabular data. *See* structured data
tangent lines, 96, 132–133
tangent planes, 96, 132, 133
tangent spaces, 96, 124, 131–135, 149
    Euclidean space and, 113
    geodesics and, 96
    linear algebra and, 133
    nonselected predictors, 135
    redundant predictors, 135
    selected predictors, 135

targets. *See* dependent variables
TDA. *See* topological data analysis
TDAmapper package, 163–166
TDAstats package, 156–157, 161
t-distributed stochastic neighbor embedding (t-SNE), 124–125, 185–186, 189–190
test error
    curse of dimensionality, 14
    overfitting, 13
testing data
    defined, 3
    overfitting, 13
text data
    overview of, 20–21
    poetry analysis project, 179–191
    vector embeddings, 20–21
topological data analysis, 159–166
    graph filtration, 76
    Mapper algorithm, 161–166
    measurement tool validation, 159–161
    multicore approaches, 194–195
    persistent homology, 155–159
    subgroup mining, 156–159
topological dimension, 82–85
tori, 86, 168
training data, defined, 3
training error, 13
transitivity
    community mining, 60
    global network metrics, 49
    vertex metrics, 43–45
Treebank tokenizer, 183
triadic closure, 43–44, 79
triangle inequality condition, 101
TSdist package, 111
t-SNE (t-distributed stochastic neighbor embedding), 124–125, 185–186, 189, 190
Twitter, 17–18, 26, 30–31

U

UCI credit default dataset, 138–139
underfitting, 13–14

undirected networks, 19
    converting to directed, 39–40
    defined, 26
    Facebook, 17, 26
    interconnectivity of networks, 48
    networks in R, 27
unstructured data, 17–21
    image data, 18–20
    network data, 17–18
    text data, 20–21
unsupervised learning, 59
    clustering vertices, 59–64
        evaluating quality outcome of clusters, 61–62
        exploring networks with random walks, 61
        overview of, 59–60
        running clustering algorithms, 62–64
        spinglass clustering, 62
    combining with supervised learning, 3
    overview of, 3
unweighted networks, 27–28, 30–32, 34–35

V

vector embeddings, 20–21
vertex metrics, 30–47
    centrality, 30–42
        authority centrality, 35
        betweenness of vertices, 32–33
        closeness of vertices, 31
        degree of vertices, 30–31
        eigenvector centrality, 33–34
        hub centrality, 35
        Katz centrality, 35
        measuring centrality in example social network, 36–42
        PageRank centrality, 34–35
    community mining, 59–60
    defined, 47
    diversity of vertices, 42
    eccentricity of vertices, 45
    efficiency of vertices, 44–45
    Forman–Ricci curvature, 45–47

prediction
    diary entry prediction in social media, 56–58
    link prediction in social media, 58–59
    triadic closure, 43–44
vertices
    betweenness of, 32–33, 64
    closeness of, 31
    community mining, 59–64
    degree of, 30–31
    depiction of, 25
    directed and undirected networks, 26
    distance between, 28
    diversity of, 42
    eccentricity of, 45, 49
    efficiency of, 44–45
    neighboring, 28
    overview of, 25
    strength of, 30
Vietoris-Rips complexes, 82

W

walktrap algorithm, 61–64
Wasserstein distance, 93, 105–107
Watts-Strogatz graphs, 52–53, 61
    network comparison, 65–66
    persistent homology, 90–93
wave functions, 197
weighted degree (strength) of vertices, 30, 82
weighted networks
    adjacency matrices, 27
    degree of vertices, 30
    disease spread tracking, 67–69
    diversity of vertices, 42
    eigenvector centrality, 34
    graph filtration, 76–81, 84–85
    PageRank centrality, 35
    paths and network distance, 28–29
    simplicial complexes, 84–85
    strength of vertices, 30

X

Xanadu, 197

Y

YouTube, 4

The Shape of Data is set in New Baskerville, Futura, Dogma, and TheSansMono Condensed.

RESOURCES

Visit *https://nostarch.com/shapeofdata* for errata and more information.

More no-nonsense books from **NO STARCH PRESS**®

THE BOOK OF R
A First Course in Programming and Statistics
BY TILMAN M. DAVIES
832 PP., $59.99
ISBN 978-1-59327-651-5

THE ART OF R PROGRAMMING
A Tour of Statistical Software Design
BY NORMAN MATLOFF
400 PP., $39.95
ISBN 978-1-59327-384-2

PRACTICAL DEEP LEARNING
A Python-Based Introduction
BY RONALD T. KNEUSEL
464 PP., $59.95
ISBN 978-1-7185-0074-7

NATURAL LANGUAGE PROCESSING WITH PYTHON AND SPACY
A Practical Introduction
BY YULI VASILIEV
216 PP., $39.95
ISBN 978-1-7185-0052-5

MATH FOR DEEP LEARNING
What You Need to Know to Understand Neural Networks
BY RONALD T. KNEUSEL
344 PP., $49.99
ISBN 978-1-7185-0190-4

DEEP LEARNING
A Visual Approach
BY ANDREW GLASSNER
776 PP., $99.99
ISBN 978-1-7185-0072-3

PHONE:
800.420.7240 OR
415.863.9900

EMAIL:
SALES@NOSTARCH.COM
WEB:
WWW.NOSTARCH.COM

Never before has the world relied so heavily on the Internet to stay connected and informed. That makes the Electronic Frontier Foundation's mission—to ensure that technology supports freedom, justice, and innovation for all people—more urgent than ever.

For over 30 years, EFF has fought for tech users through activism, in the courts, and by developing software to overcome obstacles to your privacy, security, and free expression. This dedication empowers all of us through darkness. With your help we can navigate toward a brighter digital future.

LEARN MORE AND JOIN EFF AT EFF.ORG/NO-STARCH-PRESS